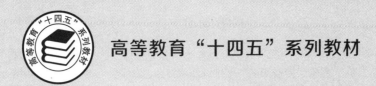

高等教育"十四五"系列教材

Android
应用开发入门实战

主　编 ◎ 陶晓霞　王立娟
副主编 ◎ 曲泰和　徐　鹏　郭　杨

电子课件

中国·武汉

内容简介

本书从初学者角度出发，以应用问题为牵引，以能力培养为目标，以企业需求为导向，以实用技能为核心，以项目案例为主线，实施"传授知识与思维训练相结合，自主学习与平台引导相结合"的学习模式。本书内容实用性强，根据项目需要讲解相关知识；同时也类似其他理论性教材，知识点讲解详细，即使没有Android基础的读者也可以理解相关内容。

本书共6章，围绕"咖啡购买App"项目开发技术和开发过程组织章节结构。第1章为Android基础入门，介绍了Android开发环境的搭建、项目简介及Android程序结构等；第2章为欢迎界面和启动界面，介绍了开发欢迎界面和启动界面的必备知识及详细步骤；第3章为登录、注册界面，介绍了如何制作登录、注册界面，如何通过网络连接外部数据库，开发登录、注册界面的详细步骤；第4章为首界面、咖啡列表和详情界面，介绍了开发首界面、咖啡列表和详情界面的必备知识，以及如何获取数据库中的数据，开发首界面、咖啡列表和详情界面的详细步骤；第5章为购物车界面，介绍了开发购物车界面的必备知识及详细步骤；第6章为我的界面，介绍了开发我的界面的必备知识，以及开发该界面的详细步骤。

本书适合作为高等院校各专业Android开发的实践教材及理论课的课后拓展教材，也适合作为计算机相关培训或技术人员自学的参考资料。

图书在版编目（CIP）数据

Android应用开发入门实战 / 陶晓霞，王立娟主编. -- 武汉：华中科技大学出版社，2024.7.
ISBN 978-7-5772-0800-8

Ⅰ．TN929.53

中国国家版本馆 CIP 数据核字第 2024HH2954 号

Android 应用开发入门实战　　　　　　　　　　　陶晓霞　王立娟　主编
Android Yingyong Kaifa Rumen Shizhan

策划编辑：	康　序
责任编辑：	朱建丽
封面设计：	岸　壳
责任校对：	王汉江
责任监印：	周治超
出版发行：	华中科技大学出版社（中国·武汉）　　电　话：（027）81321913
	武汉市东湖新技术开发区华工科技园　　邮　编：430223
录　　排：	武汉创易图文工作室
印　　刷：	武汉市洪林印务有限公司
开　　本：	787 mm×1092 mm　1/16
印　　张：	9.5
字　　数：	221千字
版　　次：	2024年7月第1版第1次印刷
定　　价：	45.00元

本书若有印装质量问题，请向出版社营销中心调换
全国免费服务热线：400-6679-118　竭诚为您服务
版权所有　侵权必究

前言

PREFACE

为什么要学习 Android？

Android 是谷歌基于 Linux 平台开发的手机及平板电脑的操作系统。安迪·鲁宾（Andy Rubin）等于 2003 年在美国成立公司，主要业务是手机软件和手机操作系统。Android 系统最初是在安迪·鲁宾引领下开发出来的，后被谷歌收购。经过多年的发展，Android 系统在全球得到了大规模推广，除了智能手机和平板电脑外，还可以用于穿戴设备、智能家居、车载系统、新能源等领域。由于 Android 的迅速发展，市场对 Android 开发人才需求日益增加。

如何使用本书？

学习本书之前，需要具备一定的 Java 和数据库的基础知识。如果是零基础读者，在使用本书时，建议从每章的相关知识开始循序渐进地学习，并且反复练习书中的案例；如果是有一定 Android 基础的读者，则可以选择感兴趣的内容进行学习。为了达到好的学习效果，建议遵循本书的章节安排，将涉及的界面一一实现。

第 1 章为 Android 基础入门，介绍了 Android 开发环境的搭建、项目简介及 Android 程序结构等。

第 2 章为欢迎界面和启动界面，介绍了开发欢迎界面和启动界面的必备知识及详细步骤。

第 3 章为登录、注册界面，介绍了如何制作登录、注册界面，如何通过网络连接外部数据库，开发登录、注册界面的详细步骤。

第 4 章为首界面、咖啡列表和详情界面，介绍了开发首界面、咖啡列表和详情界面的必备知识，以及如何获取数据库中的数据，开发首界面、咖啡列表和详情界面的详细步骤。

第 5 章为购物车界面，介绍了开发购物车界面的必备知识及详细步骤。

第 6 章为我的界面，介绍了开发我的界面的必备知识，以及开发该界面的详细步骤。

致谢

本书由陶晓霞和王立娟主编，企业工程师曲泰和、徐鹏、郭杨任副主编。全书由刘瑞杰副院长、翟悦副院长主审。本书编写过程中参考了大量的文献资料，在此感谢这些文献资料的作者。同时，感谢孙建梅老师、王小勇老师给予的建议。

为了方便教学，本书还配有电子课件等资料，任课教师可以发邮件至 hustpeiit@163.com 进行索取。

编者

2024 年 2 月

目录

CONTENTS

第1章 Android 基础入门 /001

    1.1 Android 简介 /002

    1.2 Android 开发环境搭建 /006

    1.3 项目基本情况介绍 /018

    1.4 Android 程序结构 /020

第2章 欢迎界面和启动界面 /021

    2.1 任务描述 /022

    2.2 相关知识 /022

    2.3 具体步骤 /031

第3章 登录、注册界面 /037

    3.1 任务描述 /038

    3.2 相关知识 /038

    3.3 具体步骤 /040

第4章 首界面、咖啡列表和详情界面 /073

    4.1 任务描述 /074

    4.2 相关知识 /074

    4.3 具体步骤 /076

第5章 购物车界面 /105

    5.1 任务描述 /106

    5.2 相关知识 /106

    5.3 具体步骤 /107

第6章 我的界面 /137

    6.1 任务描述 /138

    6.2 相关知识 /138

    6.3 具体步骤 /138

参考文献 /144

第 1 章
Android 基础入门

学习目标

(1)理解和掌握数据结构中的基本概念。

(2)了解 Android 发展历史、Android 体系结构。

(3)掌握 Android Studio 开发环境搭建。

(4)了解 Android 的程序结构。

1.1 Android 简介

◆ 1.1.1 Android 发展历史

Android 是基于 Linux 系统的开源操作系统，最初是由 Andy Rubin（安迪·鲁宾）创立的一个手机操作系统，2005 年被谷歌（Google）收购，并且谷歌让 Andy Rubin 继续负责 Android 项目。经过数年的研发，2007 年 11 月，谷歌向外界展示了这款名为 Android 的操作系统，并与多个硬件制造商、软件开发商组建开放手机联盟共同研发改良的 Android 系统。

2009 年 5 月，谷歌发布了 Android 1.5，该版本的 Android 界面吸引了大量开发者的目光。接下来，Android 版本更新升级非常快，几乎每隔半年就会发布一个新的版本。目前，Android 最新版本已经达到 15。Android 的每一个版本都会用一个按照 A ~ Z 开头顺序的甜品来命名，但从 Android 10 之后谷歌就改变了这一传统的命名规则，直接用数字来命名系统。Android 各版本发布时间及其代号具体如下。

2009 年 4 月 30 日，Android 1.5 Cupcake（纸杯蛋糕）正式发布。
2009 年 9 月 15 日，Android 1.6 Donut（甜甜圈）版本发布。
2009 年 10 月 26 日，Android 2.1 Éclair（松饼）版本发布。
2010 年 5 月 20 日，Android 2.2 Froyo（冻酸奶）版本发布。
2010 年 12 月 7 日，Android 2.3 Gingerbread（姜饼）版本发布。
2011 年 2 月 2 日，Android 3.0 Honeycomb（蜂巢）版本发布。
2011 年 5 月 11 日，Android 3.1 Honeycomb（蜂巢）版本发布。
2011 年 7 月 13 日，Android 3.2 Honeycomb（蜂巢）版本发布。
2011 年 10 月 19 日，Android 4.0 Ice Cream Sandwich（冰淇淋三明治）版本发布。
2012 年 6 月 28 日，Android 4.1 Jelly Bean（果冻豆）版本发布。
2012 年 10 月 30 日，Android 4.2 Jelly Bean（果冻豆）版本发布。
2013 年 7 月 25 日，Android 4.3 Jelly Bean（果冻豆）版本发布。
2013 年 9 月 4 日，Android 4.4 KitKat（奇巧巧克力）版本发布。
2014 年 10 月 15 日，Android 5.0 Lollipop（棒棒糖）版本发布。
2015 年 9 月 30 日，Android 6.0 Marshmallow（棉花糖）版本发布。
2016 年 8 月 22 日，Android 7.0 Nougat（牛轧糖）版本发布。
2017 年 8 月 22 日，Android 8.1 Oreo（奥利奥）版本发布。
2018 年 8 月 7 日，Android 9.0 Pie（派）版本发布。
2019 年 9 月 4 日，Android 10 版本发布。
2020 年 9 月 9 日，Android 11 版本发布。
2021 年 10 月 5 日，Android 12 版本发布。
2022 年 8 月 16 日，Android 13 版本发布。
2022 年 9 月，Android 14 版本发布。
2023 年 3 月，Android 15 版本发布。

Android 各版本的名称和图标如图 1-1 所示。

图 1-1 Android 各版本的名称和图标[①]

Android 是一个全身绿色的机器人，由 Ascender 公司设计，其中的文本使用了由 Ascender 公司专门制作的称之为 "Google Droid" 字体。Android 的图标代表国际性的、开源的 Android，没有借鉴任何文化角色，Android 图标如图 1-2 所示。

图 1-2 Android 图标

---

① Android 10 开始采用版本号为其名称，Android 12 ～ Android 15 没有特殊图标。

1.1.2 Android 体系结构

Android 系统采用分层架构，由高到低分为 4 层，依次是应用程序（Applications）层、应用程序框架（Application Framework）层、核心类库（Libraries）和 Linux 内核（Linux Kernel），如图 1-3 所示。

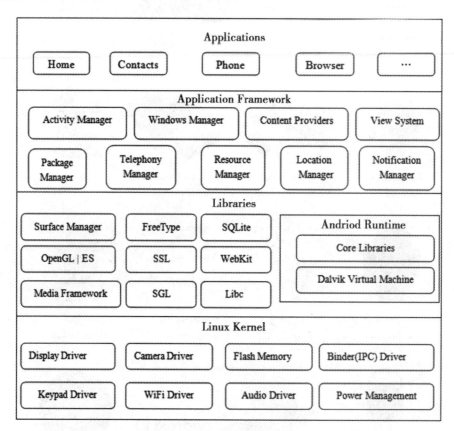

图 1-3 Android 体系结构

1. 应用程序层

应用程序层是一个核心应用程序的集合，所有安装在手机上的应用程序都属于这一层，如手机系统自带的通讯录、短信、浏览器等，或者从 Google Play 上下载的应用程序等都属于应用程序层。

2. 应用程序框架层

应用程序框架层主要提供了构建应用程序时用到的各种类库。Android 自带的应用或者从商城中下载的应用都是开发人员使用这些类库开发完成的。应用程序框架层中包含了众多的组件，部分组件介绍如下。

活动管理器（Activity Manager）：Activity 是 Android 应用程序中的基本组件，所有可运行的程序都继承 Activity 类，此类将接受 Android 操作系统的管理，此类也有自己的生命周期控制方法。

窗口管理器（Window Manager）：负责整个系统的窗口管理，可以控制窗口的打开、关闭、隐藏等操作。

内容提供者（Content Providers）：实现多个应用程序间的数据共享操作。

视图系统（View System）：用户构建应用程序的显示界面，如文本组件、按钮组件、列表显示等。

通知管理器（Notification Manager）：对手机顶部状态栏的提示消息进行管理，如短信提示、电量提示等。

资源管理器（Resource Manger）：提供访问非代码的资源，如国际化文字显示、图形界面和布局文件资源等。

位置管理器（Location Manager）：Google 提供的地图管理程序，可以为用户提供 GPS 导航功能。

3. 核心类库

核心类库中包含了系统库及 Android 运行时库（Android Runtime）。

系统库：通过 C/C++ 库来为 Android 系统提供主要的特性支持，如 OpenGL | ES 库提供了 3D 绘图的支持，WebKit 库提供了浏览器内核的支持。

Android 运行时库：主要提供了一些核心库（Core Libraries），能够允许开发者使用 Java 来编写 Android 应用程序。另外，Android 运行时库中还包含了 Dalvik 虚拟机（Dalvik Virtual Machine），它使得每一个 Android 应用程序都能运行在独立的进程当中，并且拥有一个自己的 Dalvik 虚拟机实例，相较于 Java 虚拟机，Dalvik 虚拟机是专门为移动设备定制的，它针对手机内存、CPU 性能等做了优化处理。

4. Linux 内核

Linux 内核为 Android 设备的各种硬件提供底层的驱动，如显示驱动、音频驱动、照相机驱动、蓝牙驱动、电源管理驱动等。

◆ 1.1.3 Dalvik 虚拟机

Android 应用程序的主要开发语言是 Java，它通过 Dalvik 虚拟机来运行 Java 程序。Dalvik 是谷歌设计的用于 Android 平台的虚拟机，其指令集基于寄存器架构，执行其特有的 Dex 文件来完成对象生命周期管理、堆栈管理、线程管理、安全异常管理、垃圾回收等。

每一个 Android 应用在底层都会对应一个独立的 Dalvik 虚拟机实例，其代码在虚拟机的解释下得以执行，具体过程如图 1-4 所示。

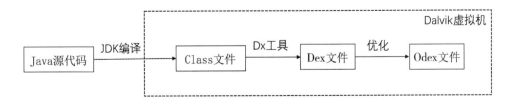

图 1-4　Dalvik 虚拟机编译文件过程

Java 源文件经过 JDK（Java Develpoment Kit）编译成 Class 文件之后，Dalvik 虚拟机中的 Dx 工具将部分（但不是全部）Class 文件转换成 Dex 文件（Dex 文件包含多个类）。Dex 文件相比 Jar 文件更加紧凑，但是为了在运行过程中进一步提高性能，Dex 文件还会被进一步优化为 Odex 文件。

需要注意的是，每个 Android 程序都运行在一个 Dalvik 虚拟机实例中，而每一个 Dalvik 虚拟机实例都是一个独立的进程空间，每个进程之间可以通信。Dalvik 虚拟机的线程机制、内存分配和管理等都是依赖底层操作系统实现的，这里不做详解，感兴趣的可以自行研究。

1.2　Android 开发环境搭建

开发 Android 应用程序需要先搭建开发环境，起初 Android 是使用 Eclipse 作为开发工具的，但在 2015 年底，谷歌声明不再对 Eclipse 提供支持服务，Android Studio 将全面取代 Eclipse 开发环境。本节将针对 Android Studio 开发工具的环境搭建进行讲解。

◆ 1.2.1　安装 Android Studio

Android Studio 是谷歌为 Android 提供的一个官方 IDE（Integrated Development Environment）工具，它集成了 Android 所需的开发工具。需要注意的是，Android Studio 对安装环境有一定的要求，其中 JDK 的最低版本为 1.7，系统空闲内存至少为 2GB，系统用户名文件夹名称为英文。

1.Android Studio 的下载

Android Studio 安装包可以从官网上下载，官网网址为：https://developer.android.google.cn/studio/archive 。这里我们以 Windows 系统为例，下载历史版本 Android Studio 3.5.2，如图 1-5 所示。读者也可以自行下载最新版本。

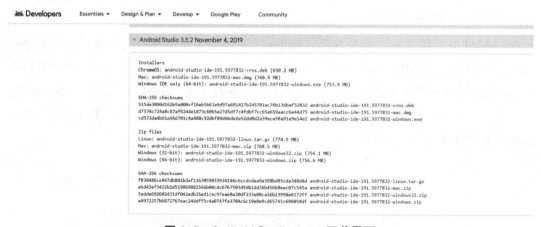

图 1-5　Android Studio 3.5.2 下载界面

2.Android Studio 的安装

成功下载 Android Studio 的安装包后，双击 Setup.exe 文件，进入 Welcome to Android

Studio Setup 窗口，如图 1-6 所示。

图 1-6　Welcome to Android Studio Setup 窗口

在图 1-6 中，单击【Next】按钮，进入 Choose Components 窗口，如图 1-7 所示。

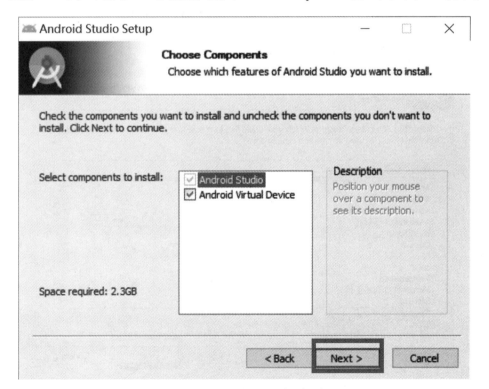

图 1-7　Choose Components 窗口

在图 1-7 中，单击【Next】按钮，进入 Configuration Settings 窗口，如图 1-8 所示。

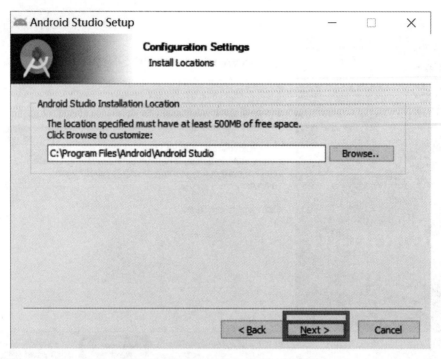

图 1-8 Configuration Settings 窗口

在图 1-8 中，可以单击【Browse】按钮来更改安装路径，我们选择不更改安装路径，使用系统默认的安装路径。然后单击【Next】按钮进入 Choose Start Menu Folder 窗口（见图 1-9），该窗口用于设置在【开始】菜单中显示的文件夹名称。

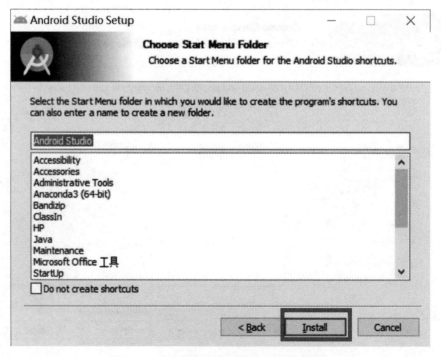

图 1-9 Choose Start Menu Folder 窗口

在图 1-9 中，单击【Install】按钮进入 Installing 窗口后开始安装，如图 1-10 所示。

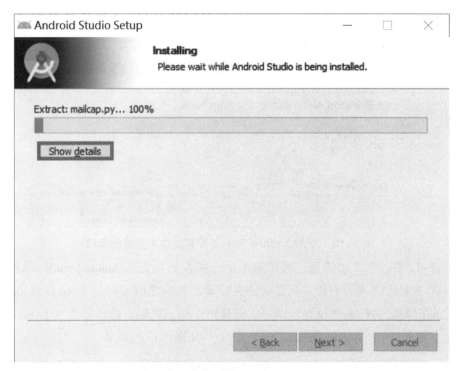

图 1-10　Installing 窗口

安装完成后，单击【Next】按钮进入 Completing Android Studio Setup 窗口，如图 1-11 所示。

图 1-11　Completing Android Studio Setup 窗口

在图 1-11 中，单击【Finish】按钮，至此 Android Studio 的安装就已经全部完成了。

3.Android Studio 的配置

如果我们在图 1-11 中勾选了【Start Android Studio】选项，安装完成之后 Android Studio 会自动启动，并弹出一个导入 Android Studio 配置文件夹位置的窗口，如图 1-12 所示。

图 1-12　导入 Android Studio 配置文件夹位置的窗口

在图 1-12 中，有 2 个选项，其中第 1 个选项表示自定义 Android Studio 配置文件夹的位置，第 2 个选项表示不导入配置文件夹位置。如果之前安装过 Android Studio，并且想导入之前的配置文件夹的位置，则可以选择第 1 项，否则，选择第 2 项【Do not import settings】，然后进入 Android Studio 的开启窗口，如图 1-13 所示。

图 1-13　Android Studio 的开启窗口

在图 1-13 中，进度完成之后，会弹出 Android Studio First Run 窗口，弹出该窗口的原因是第一次安装 Android Studio，启动后会检测到默认安装的文件夹中没有 SDK，如果单击窗口中的【Setup Proxy】按钮，则会立即在线下载 SDK。如果单击【Cancel】按钮，则暂时不下载 SDK，稍后再下载或者导入提前下载好的 SDK。此处，单击【Setup Proxy】按钮下载 SDK，由于在线下载比较慢，需要耐心等待。

下载完成后，弹出 Android Studio 的欢迎窗口，如图 1-14 所示。

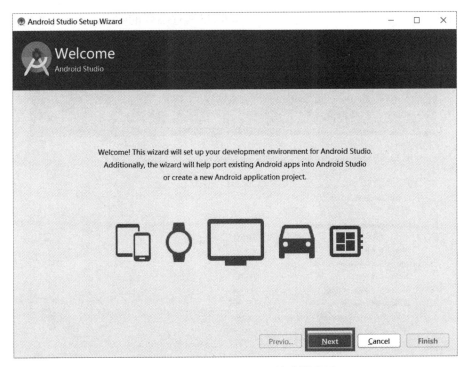

图 1-14　Android Studio 的欢迎窗口

在图 1-14 中，单击【Next】按钮进入 Install Type 窗口，如图 1-15 所示。

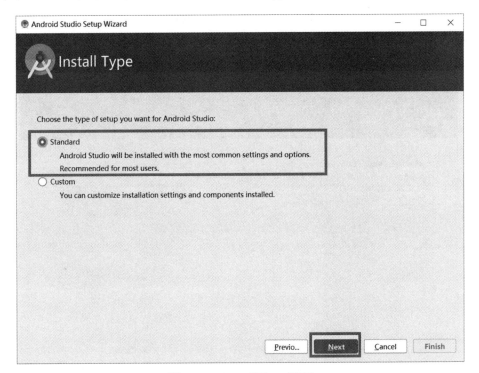

图 1-15　Install Type 窗口

在图 1-15 中，包含【Standard】和【Custom】两个选项，分别表示安装 Android Studio 的标准设置与自定义设置。如果选择【Standard】选项，则程序会默认安装很多配置，满足基本的开发要求。如果选择【Custom】选项，则需要手动进行配置安装。此处推荐选择【Standard】选项，默认安装好开发 Android 程序需要的配置。单击【Next】按钮进入 Verify Settings 窗口，如图 1-16 所示。

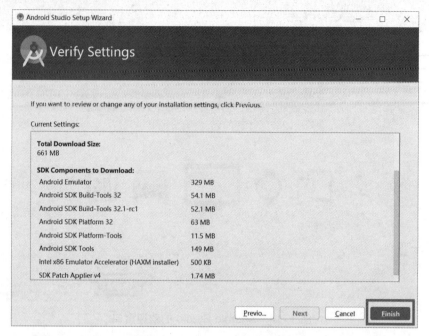

图 1-16　Verify Settings 窗口

在图 1-16 中，可以看到需要下载的 SDK 组件。单击【Finish】按钮进入 Downloading Components 窗口，如图 1-17 所示。

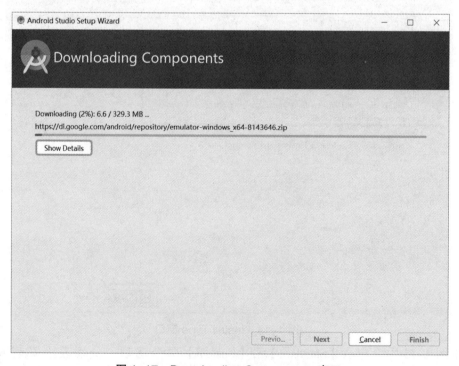

图 1-17　Downloading Components 窗口

下载完成后，会显示 Downloading Components 完成窗口，如图 1-18 所示。

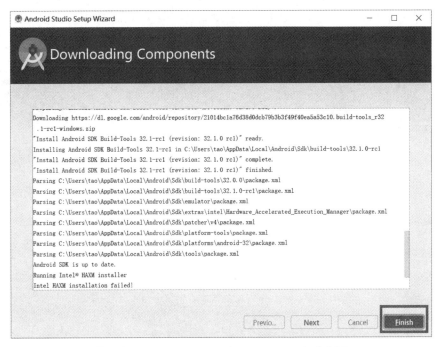

图 1-18　Downloading Components 完成窗口

在图 1-18 中，单击【Finish】按钮，进入 Welcome to Android Studio 窗口，如图 1-19 所示。

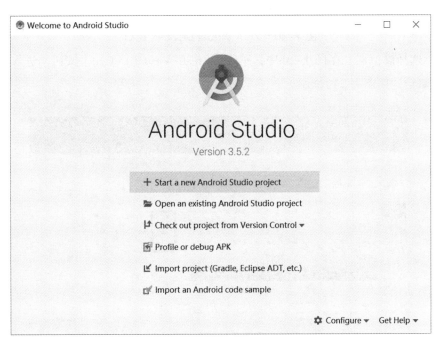

图 1-19　Weclome to Android Studio 窗口

◆ 1.2.2　创建模拟器

Android 程序可以在手机或平板电脑等物理设备上运行，当运行程序时，没有相应屏幕尺寸的物理设备时，可以使用 Android 模拟器代替。模拟器是一个可以运行在计算机上的虚拟设备。在模拟器上可预览和测试 Android 应用程序。创建模拟器的步骤如下。

（1）单击【AVD Manager】按钮。在创建完第一个 Android 程序后，在 Android Studio 中，单击导航栏中的图标会弹出 Your Virtual Devices 窗口，如图 1-20 所示。

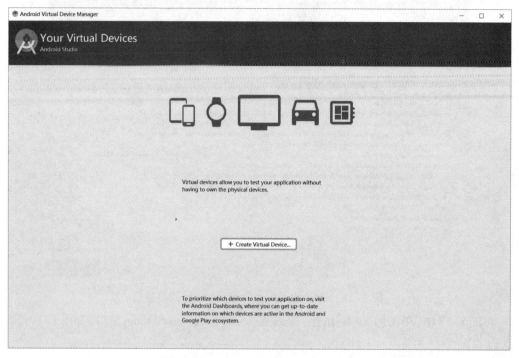

图 1-20　Your Virtual Devices 窗口

（2）选择模拟器。在图 1-20 中，单击【Create Virtual Device】按钮，进入选择模拟器设备的窗口，如图 1-21 所示。

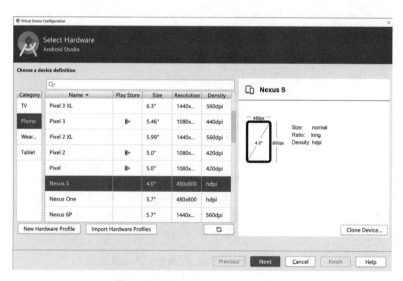

图 1-21　Select Hardware 窗口

（3）下载 SDK System Image。在图 1-21 中，其左侧是设备类型，中间对应的是设备的名称、尺寸、分辨率、密度等信息，右侧是设备的预览图。这里，我们选择【Phone】->【Nexus S】（此选项可以根据自己需求选择不同分辨率的模拟器），单击【Next】按钮进入 System Image 窗口，如图 1-22 所示。

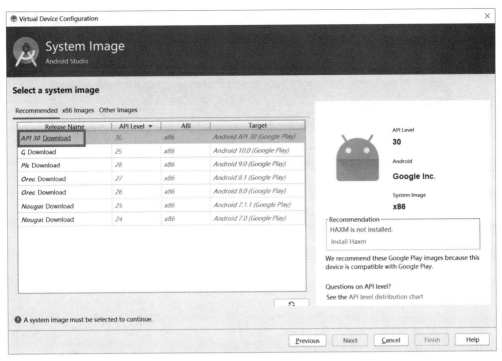

图 1-22　System Image 窗口

在图 1-22 中，左侧为推荐的 Android 系统镜像，右侧为选中的 Android 系统镜像对应的图标。此处我们选择 10.0 的系统版本进行下载，单击 Q 后面的【Downloading】，进入 License Agreement 窗口，如图 1-23 所示。

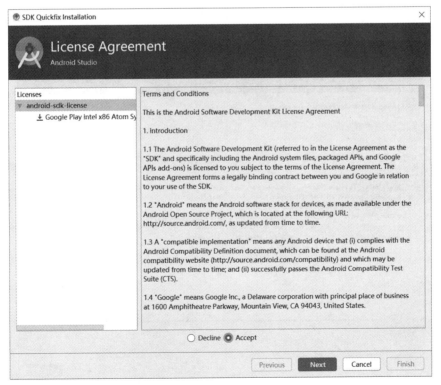

图 1-23　License Agreement 窗口

在图 1-23 中，选中【Accept】选项以接收窗口中显示的信息，单击【Next】按钮进入 Component Installer 窗口，如图 1-24 所示。

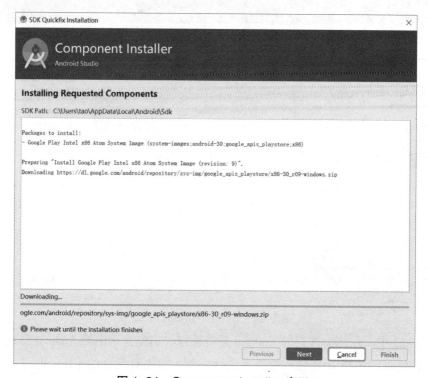

图 1-24　Component Installer 窗口

Component Installer 下载完成窗口如图 1-25 所示。

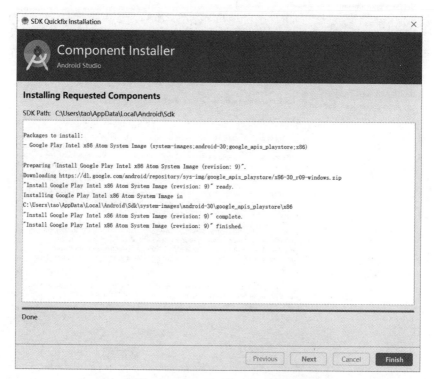

图 1-25　Component Installer 下载完成窗口

（4）创建模拟设备。在图 1-25 中，单击【Finish】按钮，弹出 System Iamge 下载完成窗口，如图 1-26 所示。

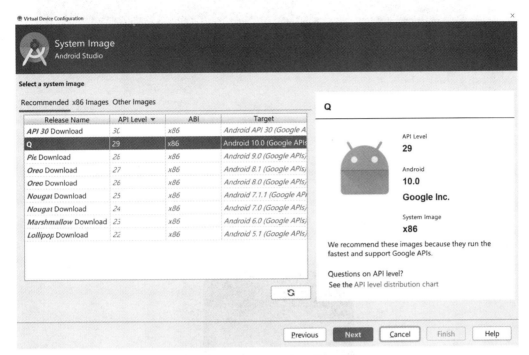

图 1-26　System Image 下载完成窗口

在图 1-26 中，单击【Next】按钮，弹出 Android Virtual Device 窗口，如图 1-27 所示。

图 1-27　Android Virtual Device 窗口

在图 1-27 中，单击【Finish】按钮，完成模拟器的创建。此时在 Your Virtual Devices 窗口中会显示创建完成的模拟器，如图 1-28 所示。

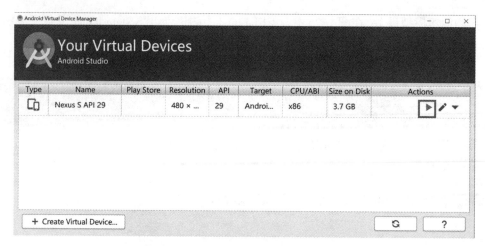

图 1-28　Your Virtual Devices 窗口

（5）打开模拟器。单击图 1-28 中的启动按钮，启动模拟器，模拟器窗口如图 1-29 所示。

图 1-29　模拟器窗口

1.3　项目基本情况介绍

◆ 1.3.1　项目简介

商品类 App 一般以用户购买商品流程分为商品浏览、商品详情查看、加入购物车、商品购买、生成订单、订单查看等流程。客户端功能主要在手机端进行显示。本实战案例简化了商品类 App 功能，实现了用户可以通过手机 App 进行个人账号管理、登录、商品信息查看、商品详情查看等功能，完成了通过手机购买咖啡的需求。

1.3.2 项目开发环境

该项目用到的软件环境如下。

Android Studio 版本：Android Studio 3.5.2。
Gradle 版本：gradle-5.4.1-all。
JDK：1.8.0。
AVD：WVGA（Nexus S）API 29（分辨率 480×800）。
Navicat Lite 版本：9.0.12。
Eclipse 版本：Eclipse Java EE IDE for Web Developers。
Tomcat 版本：8.0.5。

1.3.3 项目实现方案

系统大量的信息存储到远程服务器上，远程服务器采用 MySQL 数据库，每当手机端需要从数据库中提取信息或更新数据库信息时，首先要向服务器发送请求，服务器接收到请求后到后台数据库提取或更新相应信息，然后把操作结果返回给手机端的客户端。

服务器端与客户端采用 AsyncHttpClient 方式通信，数据格式采用 JSON。手机端存储少量信息，采用 SharedPreferences 存储实现登录状态保持的 sessionid 和判断用户是否为第一次登录的信息，网站用到的一些商品图片存储在 Tomcat 中。

1.3.4 新建项目

1.2 节已经搭建好了开发环境，接下来使用 Android Studio 工具开发第一个 Android 程序，具体步骤如下。

第 1 步：新建项目，项目名称为"CoffeeDemo"，设置应用程序支持的最低 SDK 版本为 4.3，如图 1-30 所示。

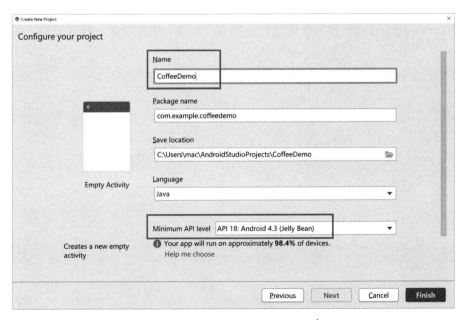

图 1-30　Configure your project 窗口

在这里，需要填写的信息主要有【Name】【Package name】【Save location】【Language】和【Minum API level】。【Name】是应用程序的名称。【Packgage name】是包名，输入应用程序名称后，包名会自动生成，当然也可以对其进行修改。【Save location】是应用程序项目源码存放的地址，可以点击右侧的图标进行修改。【Language】是开发语言，有 Java 和 Kotlin 两种可选。【Minimum API level】是设置应用程序支持的最低 SDK 版本。

第 2 步：App 实现全屏显示，分 2 小步。

（1）找到 res/values/styles.xml 文件，在 <resources> 节点下添加如下代码：

```
1    <style name="CustomTheme" parent="@style/AppTheme">
2        <item name="android:windowNoTitle">true</item>
3    </style>
```

（2）打开 manifests/AndroidManifest.xml 文件，修改 <application> 节点的 "android:theme" 属性值为 "CustomTheme"，代码如下：

```
1    <application
2        android:theme="@style/CustomTheme">
```

1.4 Android 程序结构

创建 Android 应用程序后，Android Studio 就会为其构建基本结构，开发人员可以在此结构上开发应用程序。接下来，我们以 1.3 节创建的应用程序 CoffeeDemo 为例，介绍 Android 程序的主要组成结构。CoffeeDemo 程序结构如图 1-30 所示，具体介绍如下。

（1）AndroidManifext.xml：为整个应用程序的配置文件，在该文件中配置程序所需的权限和注册程序中用到的组件。

（2）Java 文件夹：用于存放程序的代码文件。

（3）res：用于存放程序的资源文件，如布局资源、图片资源、样式资源、字符串资源等。

（4）build.gradle：为程序的 gradle 构建脚本。

第 2 章
欢迎界面和启动界面

学习目标

通过欢迎界面和启动界面的实战学习,学生能够分析程序的基本功能,利用所学知识完成界面搭建,具体目标如下。

(1)理解布局容器 <LinearLayout>、常用控件 <Button> 的作用及其主要属性。

(2)理解 Intent,并能运用 Intent 完成界面跳转。

(3)理解线程的概念,并能运用自定义线程。

(4)理解共享参数存储,学会使用共享参数存储。

2.1 任务描述

启动 App 时，首先启动欢迎界面（见图 2-1），停留 5 s 后，自动跳转到启动界面（见图 2-2）。

图 2-1　欢迎界面效果图

图 2-2　启动界面效果图

2.2 相关知识

◆ 2.2.1　布局概述

应用界面包含用户可查看并与之互动的所有内容。Android 提供丰富多样的预构建界面组件，如结构化布局对象和界面控件，我们可以利用这些组件为应用构建图形界面。Android 还提供其他界面模块，用于构建特殊界面，如窗口、通知和菜单。

1. 布局

布局定义了应用中的界面结构，布局中的所有元素均使用 View 和 ViewGroup 对象的层次结构进行构建。View 通常用于绘制用户可以看到并与之交互的内容，如 Button、EditText、TextView 等。ViewGroup 则是不可见的容器，如 LinearLayout、ConstrainLayout 等，用于定义 View 和其他 ViewGroup 对象的布局结构，如图 2-3 所示。

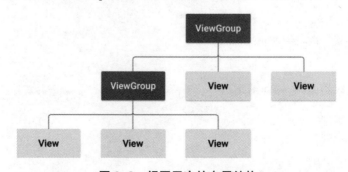

图 2-3　视图层次的布局结构

通常，可以通过两种方式声明布局。

在 XML 中声明界面元素，编写布局文件，在 XML 中创建包含一系列的嵌套元素，但每个布局文件必须只包含一个根元素，并且该元素必须是 ViewGroup 对象。

在运行时实例化布局元素，即在 Java 代码中编写布局。在 Android 中所有布局和控件的对象都可以通过 new 关键字创建，将创建的 View 控件添加到 ViewGroup 布局中，实现 View 控件在布局界面中显示。

2. 常用的共有属性

每个 View 对象和 ViewGroup 对象均支持自己的各种 XML 属性，某些属性是 View 对象的特有属性（如 TextView 支持 textSize 属性），但继承自 TextView 的 View 也具有这些属性。

而某些属性是所有 View 对象的共有属性，因为它们继承自 View 类（如 id 属性），下面介绍几个常用的共有属性。

（1）id 属性。

任何 View 对象均可拥有与之关联的整型 id，用于在结构树中对 View 对象进行唯一标识。在布局 XML 文件中，系统通常会以字符串形式在 id 属性中指定 id，这是所有 View 对象共有的 XML 属性。语法规则如下：

```
1    android:id="@+id/my_button"
```

字符串开头处的 @ 符号表示 XML 解析器应解析并展开 id 字符串的其余部分，并将其表示为 id 资源，加号表示这是一个新的资源名称，必须创建该名称并将其添加到资源（R.java 文件）中。

为了创建视图并从 Java 文件中引用它们，可以进行如下操作。

首先在布局文件中定义视图并为其分配唯一的 id，代码如下：

```
1    <Button
2        android:id="@+id/my_button"
3        android:layout_width="wrap_content"
4        android:layout_height="wrap_content"
5        android:text=" 我是按钮 "/>
```

然后可以创建视图对象的实例，并从布局文件中捕获它，代码如下：

```
1    Button myButton=(Button)findViewById(R.id.my_button);
```

（2）宽度和高度。

所有的视图都要包含宽度和高度，因此每个视图都必须被定义，语法规则如下：

```
1    android:layout_width="XX dp | match_parent | wrap_content"
2    android:layout_height="XX dp | match_parent | wrap_content"
```

其值可以是具体的尺寸，如 30dp，也可以是系统定义的值，介绍如下。

match_parent：表示该视图的宽度和高度与父元素相同。

wrap_conent：表示该视图的宽度和高度为内容所需要的尺寸。

（3）外边距和内边距。

外边距用于设置当前视图与其他视图之间的距离，其属性值为具体的尺寸，如 30dp，其语法格式如下：

```
1    android:layout_margin="30dp"
```

与之相似的还有 android:layout_marginTop、android:layout_marginRight、android:layout_marginBottom、android:layout_marginLeft 属性，分别用于设置当前视图与其他视图的上、右、下、左边界的距离。

内边距用于设置当前视图的内容与该视图边界之间的距离，其值为具体的尺寸，如 10dp，语法格式如下：

```
1    android:padding="10dp"
```

与之相似的还有 android：paddingTop、android：paddingRight、andriod：paddingBottom、android:paddingLeft 属性，分别用于设置当前视图的内容与该视图之间的上、右、下、左边界的距离。

◆ 2.2.2 LinearLayout 线性布局

视图布局常见的有线性布局、相对布局、网格布局等，这里介绍最常用的 LinearLayout（线性布局），该布局中的控件以水平或者垂直的方式排列。在 XML 布局文件中使用线性布局的语法格式如下：

```
1    <LinearLayout
2        属性 =" 属性值 "
3        ... >
4    </LinearLayout>
```

其属性介绍如下。

（1）android:orientation：方向属性，用于控制控件的排列方向，属性值有两种，分别是 "horizontal" 和 "vertical"。其中 "horizontal" 表示 LinearLayout 线性布局中的控件以水平方式排列，即控件依书写顺序从上到下依次排列。"vertical" 表示 LinearLayout 线性布局中的控件以垂直方式排列，即控件依书写顺序从左到右垂直排列。

（2）android:layout_weight：权重属性，通过设置该属性，可以使布局内的控件按照权重比显示大小，在进行屏幕适配时起到重要作用。

◆ 2.2.3 Button 组件

几乎每一个 Android 应用都是通过界面组件与用户交互的，Android 提供了非常多的组件，通过这些组件可以进行界面的开发。

Button 组件表示按钮，它可以显示文本，还可以显示图片，主要用于响应用户的点击事件，基本语法格式如下：

```
1    <Button
2        android:layout_width=" 设置宽度 "
3        android:layout_height=" 设置高度 "
4        android:text=" 设置文本内容 "
5        android:background=" 设置背景 "
6        … />
```

其属性介绍如下。

（1）android:layout_width：设置控件的宽度，属性值有三种，match_parent、wrap_content 和固定值，"match_parent" 表示宽度与父元素相同，"wrap_content" 表示宽度由自身内容决定，固定值表示宽度固定不变。

（2）android:layout_height：设置控件的高度，属性值有三种，与 android:layout_width 的属性值相同，这里不再赘述。需要注意的是，android:layout_width 和 android:layout_height 是所有视图 View 都具有的属性，即布局容器和控件都需要设置宽度和高度属性。

（3）android:text：设置要显示的文本内容。

（4）android:background：设置视图的背景，属性值可以是图片，也可以是颜色值，这也是所有视图 View 都具有的属性。

◆ 2.2.4 Activity 类简介

Activity 类是 Android 应用的关键组件，而 Activity 类的启动是学习 Android 必须要掌握的部分。在编程规范中，应该通过 main（）方法启动，而 Android 系统与此不同，它会调用生命周期方法来启动 Activity 实例中的代码。

Activity 类提供窗口供应用在其中绘制界面，此窗口通常会填满屏幕，但也可能比屏幕小，并浮动在其他窗口上面。通常，一个 Activity 类实现应用中的一个屏幕。大多数应用包含多个屏幕，这意味着它们包含多个 Activity 类。

当用户浏览、退出或返回到应用程序时，应用程序中的 Activity 实例就会在其生命周期的不同状态之间进行转换。Activity 类会提供许多回调，这些回调会让 Activity 实例知晓某个状态已经变更：系统正在创建、停止或恢复某个 Activity 实例，或者正在销毁该 Activity 实例所在的进程。

1. Activity 实例的生命周期

Activity 实例的生命周期指的是 Activity 实例从启动到销毁的整个过程，这个过程（见图 2-4）大致分为六种状态，分别是启动状态、开始状态、恢复状态、暂停状态、停止状态和销毁状态，下面详细介绍这六种状态。

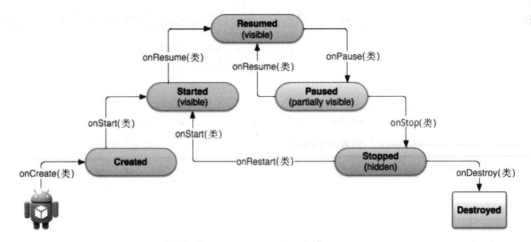

图 2-4　Activty 类生命周期简化图

（1）启动状态（Created）和开始状态（Started）：这两个状态都很短暂，是瞬时状态，Activity 类会迅速从这两个状态切换到下一个状态。

（2）恢复状态（resumed）：该 Activity 类位于前台，用户可以与它交互。

（3）暂停状态（paused）：该 Activity 类被另一个活动部分覆盖，如另一个 Activity 类在前台是半透明或者不覆盖整个屏幕。暂停状态不接受用户输入，也不执行任何代码，里面包含所有状态信息，会被附加到窗体，但是在内存极度不足情况下会被系统销毁。

（4）停止状态（stopped）：该 Activity 类被完全隐藏并对用户不可见，它被认为是在后台中，停止时 Activity 实例及其多种状态都被保留，但它不能执行任何代码，不会被附加到窗体，系统需要时被系统销毁。

（5）销毁状态（destoryed）：当进入销毁状态时，Activity 类将被清理出内存。

2.Activity 类的生命周期方法

为了在 Activity 类生命周期的各个阶段之间进行转换，Activity 类提供了六个核心回调：onCreate（）、onStart（）、onResume（）、onPause（）、onStop（）和 onDestory（）。当 Activity 进入新状态时，系统会调用其中每个回调。

1）onCreate（）方法

必须要实现该回调函数，它会在系统首次创建 Activity 类时触发。创建 Activity 类后其就进入启动状态，在 onCreate（）方法中，需要执行基本的应用启动，如初始化某些类的变量、数据绑定等，如下代码为显示执行 Activity 类某些基本设置的一些代码，如声明界面、定义成员变量。

```
1    public void onCreate(Bundle savedInstanceState) {
2        super.onCreate(savedInstanceState);
3        setContentView(R.layout.main_activity);
4    }
```

onCreate（）方法用于执行 Activity 类某些基本设置。

代码 2 行，调用 onCreate（）方法的父类可以完成 Activity 类的创建。

代码 3 行，系统通过将文件的资源 ID "R.layout.main_activity" 传递给 setContentView（）方法来指定要加载的 XML 布局文件。

onCreate（）方法执行完成后，Activity 类就进入开始状态，系统会相继调用 onStart（）方法和 onResume（）方法。

2）onStart（）方法

当 Activity 进入开始状态时，系统会调用此回调。onStart（）方法使 Activity 类对用户可见，因为应用会为 Activity 类进入前台并为支持互动做准备。例如，应用通过此方法来初始化维护界面的代码。

onStart（）方法会非常快速地完成，并且与启动状态一样，Activity 类不会一直处于开始状态。一旦此回调结束，Activity 类便会进入恢复状态，系统将会调用 onResume（）方法。

3）onResume（）方法

Activity 类会在进入恢复状态时来到前台，然后系统调用 onResume（）方法来进行回调。这是应用与用户互动的状态。应用会一直保持这种状态，直到某种事件发生，让焦点远离应用。此类事件包括接到来电、用户导航到另一个 Activity 类，或设备屏幕关闭。

当 Activity 类进入恢复状态时，与 Activity 类生命周期相关联的所有生命感知型组件都会收到 ON_RESUME 事件。这时，生命周期组件可以启用在组件可见且位于前台时需要运行的任何功能，如启动照相机。

当发生中断事件时，Activity 类进入暂停状态，系统调用 onPause（）方法来进行回调。

4）onPause（）方法

系统将此方法视为用户将要离开 Activity 类的第一个标志，此方法表示 Activity 类不再位于前台（尽管在用户处于多窗口模式时 Activity 类仍然可见）。Activity 类进入此状态的原因有很多，如 onResume（）方法部分所述，某个事件会中断应用执行，这是最常见的情况。

有半透明的 Activity 类（如窗口）处于开启状态，只要 Activity 类仍然部分可见但并未获得焦点，它便会一直处于暂停状态。

还可以使用 onPause（）方法释放系统资源、传感器手柄等。当 Activity 类处于暂停状态时，它有两种选择：

如果 Activity 类变为完全不可见，系统会调用 onStop（）方法。

如果 Activity 类获得焦点，系统会调用 onResume（）方法。

5）onStop（）方法

如果 Activity 类不再对用户可见，说明系统已进入停止状态，因此系统会调用 onStop（）方法来进行回调。例如，当新启动的 Activity 类覆盖整个屏幕时，可能会发生这种情况。

在 onStop（）方法中，应用应释放或调整在应用对用户不可见时的无用资源。例如，应用可以暂停动画效果。在进入停止状态后，Activity 类有以下两种选择。

（1）要么返回与用户互动，依次调用 onRestart（）方法、onStart（）方法和 onResume（）方法。

（2）要么结束运行并消失，调用 onDestory（）方法。

6）onDestory（）方法

销毁 Activity 类之前，系统会先调用 onDestory（）方法。onDestory（）方法回调应释放先前的回调尚未释放的所有资源。

3.Activity 类状态变更

用户触发和系统触发的不同事件都会导致 Activity 类从一个状态转换到另一个状态。下面详细介绍发生此类转换的一些常见情况，以及如何处理这些转换。

1）Activity 类或窗口显示在前台

如果有新的 Activity 类或窗口出现在前台，并且局部覆盖了正在进行的 Activity 类，则被覆盖的 Activity 类就会失去焦点并进入暂停状态。然后，系统会调用 onPause（）方法。

如果有新的 Activity 类或窗口出现在前台，夺取了焦点且完全覆盖了正在进行的 Activity 类，则被覆盖的 Activity 类就会失去焦点并进入停止状态。然后，系统会快速接连调用 onPause（）和 onStop（）方法。

2）用户点击【返回】按钮。

如果 Activity 类位于前台，并且用户按了【返回】按钮，Activity 类依次经历 onPause（）方法、onStop（）方法、onDestory（）方法的回调。该 Activity 类不仅会被销毁，还会被移出堆栈。

◆ 2.2.5　Intent 类简介

在 Android 中，一般应用程序都是由多个界面构成的，如果用户需要从一个界面切换到另一个界面，则必须使用 Intent 类来进行切换。不仅如此，Activity、Service 和 BroadcastReceiver 这三种核心组件都需要使用 Intent 类进行触发。

Intent 类被称为意图，是程序中各组件进行交互的一种重要方式，它不仅可以指定当前组件要执行的动作，还可以在不同组件之间进行数据传递。根据开启目标组件的方式不同，Intent 类被分为显式 Intent 类和隐式 Intent 类。

1. 显式 Intent 类

显式 Intent 类指的是直接指定目标组件的类名，常被用于在自定义的界面之间进行跳转，语法格式如下：

```
1    Intent intent=new Intent(当前Activity类名.this,跳转到的Activity
     类名.class);
2    startActivity(intent);
```

其中，代码 1 行实例化 Intent 类的对象时传入了 2 个参数，第 1 个参数表示当前的 Activity 类，第 2 个参数表示要跳转到的目标 Activity 类。

2. 隐式 Intent 类

隐式 Intent 类不会明确指出需要激活的目标组件，它被广泛地应用在不同应用之间传递信息。Android 系统会使用 IntentFilter 来匹配相应的组件，匹配的属性主要包括以下三个。

（1）action：为 Intent 对象要完成的动作。

（2）data：为 Intent 对象中传递的数据。

（3）category：为 action 添加的额外信息。

例如，要切换到拨号界面，代码如下：

```
1   Intent intent=new Intent();
2   intent.setAction(Intent.ACTION_DIAL);
3   intent.setData(Uri.parse("tel:15966668888"));
4   startActivity(intent);
```

其中，代码 3 行的 setData（）方法可以传递电话号给拨号界面，但需要是 Uri 格式。

◆ 2.2.6　自定义线程

在 Android 中，启动应用时，Android 系统会为该应用创建一个称为"main"（主线程）的执行线程。此线程非常重要，因为其负责将事件分派给相应的界面组件，其中包括绘图事件，应用与 Android 界面工具包组件（android:widget 和 android:view 软件包的组件）也几乎都在该线程中进行交互。

但是当应用执行繁重的任务以响应用户交互时，除非正确实现应用，否则这种单线程模式可能会导致性能低下。具体地讲，如果界面线程需要处理所有任务，则执行耗时较长的操作（如网络访问或数据库查询）将会阻塞整个界面线程。一旦被阻塞，线程将无法分配任何事件。更糟糕的是，如果界面线程被阻塞超过几秒（目前大约是 5 s），用户便会感到厌烦，引起用户不满。

因此，耗时的操作需要在自定义线程中完成。所谓自定义线程，指的是用户使用 new Thread（）方法定义的线程，语法格式如下：

```
1   public void onClick(View v) {
2       new Thread(new Runnable() {
3           public void run() {
4               // a potentially time consuming task
5           }
6       }).start();
7   }
```

◆ 2.2.7　应用数据和文件存储概览

Android 使用的文件系统类似于其他平台基于磁盘的文件系统。该系统提供了以下几种保存应用数据的选项。

应用专属存储空间：存储仅供使用的文件，可以存储的内容为存储中的专属目录或外部存储空间中的其他专属目录。使用内部存储空间的目录保存其他应用不应访问的敏感信息。

共享存储：存储的应用将与其他应用共享的文件，包括媒体、文档或其他文件。

偏好设置：以键值对形式存储私有原始数据。

数据库：结构化数据存储在 SQLite 数据库中。

根据保存的数据需要占用多少空间、数据访问需要达到怎样的可靠程度、存储哪类数据、数据是否应仅供应用使用等，选择适合的存储方式。

◆ 2.2.8 存储键值对数据

如果想要保存的相对较小键值对集合，则应该使用共享参数存储，即使用 SharedPreferences API。SharedPreferences 对象指向包含键值对的文件，并提供读写这些键值对的简单方法。

共享参数存储是通过 key/value（键值对）的形式将数据保存在 XML 文件中，该文件位于 data/data/<packagename>/shared_prefs 文件中，使用共享偏好设置的方法如下。

1. 获取共享偏好设置文件

如果要获取共享偏好设置文件，代码如下：

```
1   SharedPreferences sharedPref=
        getActivity().getSharedPreferences("data",Context.MODE_
        PRIVATE);
```

其中，getSharedPreferences（）方法需要 2 个参数，第 1 个参数是共享偏好设置文件名，可以自定义；第 2 个参数是文件的读写模式。

2. 写入共享偏好设置文件

如果需要写入共享偏好设置文件，代码如下：

```
1   SharedPreferences sharedPref=getActivity().getPreferences(Context.
    MODE_PRIVATE);
2       SharedPreferences.Editor editor=sharedPref.edit();
3       editor.putInt(getString(R.string.saved_high_score_key),
        newHighScore);
4       editor.commit();
```

3. 从共享偏好设置中读取文件

如果需要从共享偏好设置文件中检索值，请调用 getInt（）方法和 getString（）方法等，为检索值提供键，如果键不存在，则可以选择返回默认值，代码如下：

```
1   SharedPreferences sharedPref=getActivity().getPreferences(Context.
    MODE_PRIVATE);
2       int defaultValue=getResources().getInteger(R.integer.saved_high_
        score_default_key);
3       int highScore=sharedPref.getInt(getString(R.string.saved_high_
        score_key),defaultValue);
```

2.3 具体步骤

◆ 2.3.1 新建 WelcomeActivity

第 1 步：选中"com.example.coffeedemo"包，点击右键，选择【new】→【package】，打开"New Package"窗口，新建"activity"包和"entity"包，然后将 MainActivity 拖曳到【activity】包下，结果如图 2-5 所示。

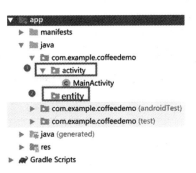

图 2-5 新建"acivity"包和"entity"包

第 2 步：选中"activity"包，点击右键，选择【new】→【Activity】→【Empty Activity】，弹出如图 2-6 所示窗口，在【Activity Name】中输入名称"WelcomeActivity"，单击【Finish】按钮完成创建，创建后欢迎界面的目录结构如图 2-7 所示。

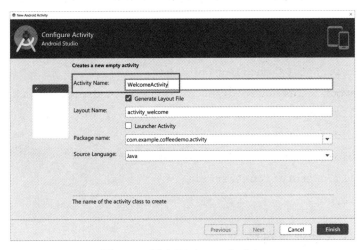

图 2-6 新建 WelcomeActivity

图 2-7 创建欢迎界面后的目录结构

第 3 步：修改配置文件 AndroidManifest.xml 文件，将 WelcomeActivity 界面作为系统的默认启动界面，manifests/AndroidManifest.xml 文件修改如图 2-8 所示。

```xml
<application
    android:allowBackup="true"
    android:icon="@mipmap/ic_launcher"
    android:label="CoffeeDemo"
    android:roundIcon="@mipmap/ic_launcher_round"
    android:supportsRtl="true"
    android:theme="@style/CustomTheme">
    <activity android:name=".activity.WelcomeActivity">
        <intent-filter>
            <action android:name="android.intent.action.MAIN" />
            <category android:name="android.intent.category.LAUNCHER" />
        </intent-filter>
    </activity>
    <activity android:name=".activity.MainActivity">
    </activity>
</application>
```

图 2-8 修改配置文件

◆ **2.3.2 搭建欢迎界面布局**

选择【res】→【layout】，打开"activity_welcome.xml"文件，根据效果图（见图 2-1）完成界面搭建。

res\layout\activity_welcome.xml 代码如下：

```
1    <?xml version="1.0" encoding="utf-8"?>
2    <LinearLayout xmlns:android="http://schemas.android.com/apk/res/android"
3        android:layout_width="match_parent"
4        android:layout_height="match_parent"
5        android:background="@drawable/welcome"
6        android:orientation="vertical">
7    </LinearLayout>
```

◆ **2.3.3 搭建启动界面布局**

选择【res】→【layout】，打开"activity_main.xml"文件，根据效果图（见图 2-2）完成界面搭建。

res\layout\activity_main.xml 代码如下：

```
1    <?xml version="1.0" encoding="utf-8"?>
2    <RelativeLayout xmlns:android="http://schemas.android.com/apk/res/android"
3        android:layout_width="match_parent"
4        android:layout_height="match_parent"
5        android:background="@drawable/bg_main">
```

```
6       <LinearLayout
7           android:layout_width="wrap_content"
8           android:layout_height="wrap_content"
9           android:orientation="horizontal"
10          android:layout_alignParentBottom="true"
11          android:layout_marginBottom="80dp"
12          android:layout_centerHorizontal="true">
13          <Button
14              android:id="@+id/btn_login"
15              android:layout_width="wrap_content"
16              android:layout_height="wrap_content"
17              android:text=" 登录 "
18              android:background="#fff"/>
19          <Button
20              android:id="@+id/btn_regist"
21              android:layout_width="wrap_content"
22              android:layout_height="wrap_content"
23              android:text=" 注册 "
24              android:background="#fff"
25              android:layout_marginLeft="40dp"/>
26      </LinearLayout>
27  </RelativeLayout>
```

◆ 2.3.4 完成欢迎界面功能逻辑实现

判断是否是第一次启动，如果是则停留 5 s 跳转到启动界面，如果不是第一次启动，则直接跳转到启动界面。

程序如何知道用户是否为第一次启动呢？利用一个变量标记用户是否为第一次登录，并将该变量以共享参数存储的方式存储到某一文件中，每次启动前需要先读取该变量并判断其值。

实现欢迎界面的关键技术有如下两个。

（1）利用布尔类型变量 isFirstRun 的值来判断 App 是否是第一次启动，将变量用共享参数存储的方式存储到设备中，每次 App 启动时，先判断 isFirstRun 的值。

（2）"在某一界面停留 n 秒，然后再跳转"是耗时的操作，耗时操作要在自定义线程中完成。

java\com\example\coffeedemo\activity\WelcomeActivity.java 代码如下：

```
1   package com.example.coffeedemo.activity;
2
3   import androidx.appcompat.app.AppCompatActivity;
4   import android.content.Intent;
```

```java
5   import android.content.SharedPreferences;
6   import android.os.Bundle;
7   import android.widget.Toast;
8   import com.example.coffeedemo.R;
9
10  public class WelcomeActivity extends AppCompatActivity {
11      protected void onCreate(Bundle savedInstanceState) {
12          super.onCreate(savedInstanceState);
13          setContentView(R.layout.activity_welcome);
14          SharedPreferences sp=getSharedPreferences("config",MODE_PRIVATE);
15          Boolean data=sp.getBoolean("isFirstRun",true);
16          if(!data){
17              Intent intent=new Intent(WelcomeActivity.this,MainActivity.class);
18              startActivity(intent);
19              finish();
20          }else{

21              new Thread(){
22                  public void run() {
23                      super.run();
24                      try {
25                          Thread.sleep(5000);
26                          SharedPreferences sp=getSharedPreferences("config",MODE_PRIVATE);
27                          SharedPreferences.Editor editor=sp.edit();
28                          editor.putBoolean("isFirstRun",false);
29                          editor.commit();
30                          Intent intent=new Intent(WelcomeActivity.this,MainActivity.class);
31                          startActivity(intent);
32                      } catch (InterruptedException e) {
33                          e.printStackTrace();
34                      }
35                  }
36              }.start();
37          }
38      }
39  }
```

代码 14、15 行，第一次启动 App 时，调用 SharedPreferences 实例的 getBoolean（String key, Boolean defValue）方法，从名为 "config" 的共享参数文件中取出值，因为没有存储过，所以显然是取不出来的，这时采用默认值 true。

代码 16 ~ 20 行，如果不是第一次启动 App，则跳转到启动界面，并关闭欢迎界面，不会显示欢迎界面。

代码 21 ~ 36 行，通过 "new Thread（）{}.start（）；" 定义线程并启动线程，重写 run（）方法。代码 25 行，调用 sleep（）方法使系统停留 5000 ms。代码 26 ~ 29 行，将共享参数文件 config 中的 "isFirstRun" 的值改为 false。代码 30 行，实例化 Intent 类，指定要跳转到的界面。代码 31 行，调用 startActivity（）方法启动 Intent 实例。

运行项目，选中 App 所在文件夹，选择菜单【run】→【run app】，等待项目自动安装到模拟器上，查看欢迎界面和启动界面的运行结果。

第 3 章

登录、注册界面

学习目标

通过登录注册界面的实战学习,学生能够分析程序的基本功能,利用所学知识完成界面搭建,具体目标如下。

(1)理解编辑框组件 \<EditText\> 的常用属性,并掌握使用编辑框组件的方法。

(2)掌握使用 AsyncHttpClient 发送请求和获取服务器响应的方法。

(3)掌握使用 session 实现登录状态保持的方法。

3.1 任务描述

启动 CoffeeDemo 后，客户端需要输入合法的手机号和密码进行登录（见图 3-1），用户输入后将手机号和密码传送到服务器端程序，服务器端程序对此进行判断，返回用户是否合法，客户端接收到服务器端返回的结果，解析结果数据，根据结果给予用户不同的响应，如果是合法用户，则登录成功，进入到主界面，否则，登录失败，提示错误信息。

如果没有用户名和密码，则需要先注册（见图 3-2），注册后，将手机号和密码传送到服务器端程序，服务器端程序将数据存入 MySQL 数据库，返回注册成功结果。

图 3-1　登录界面效果图

图 3-2　注册界面效果图

3.2 相关知识

◆ 3.2.1 编辑框组件

EditText 是编辑框，它是 TextView 的子类，用户可以用该组件输入、编辑信息。继承了所有 TextView 的属性外，EditText 还有一些其他的常用属性，如可以设置输入格式，代码如下：

```
1    <EditText
2        android:layout_width=" 设置宽度 "
3        android:layout_height=" 设置高度 "
4        android:hint=" 设置提示文本信息 "
5        android:password="true|false"
6        android:phoneNumber="true|false"
7        android:textColorHint=" 设置提示文本信息的颜色 "
8        …/>
```

其属性介绍如下：

（1）android:layout_width 和 anroid:layout_height 属性的含义已经在第 2 章中进行了介绍，这里不再赘述。

（2）android:hint：设置编辑框为空时显示的提示文本信息。

（3）android:password：当属性值为 true 时，编辑框中的内容显示为"."。

（4）android:phontNumber：当属性值为 true 时，编辑框中的内容只能是数字。

（5）android:textColorHint：设置提示文本信息的颜色。

◆ 3.2.2 网络连接

在实际开发过程中，绝大多数 App 都需要与服务器进行数据交互，也就是访问网络，此时就需要用到 Android 提供的 API。Android 提供了多种实现网络通信的方式，有 Android 原生的 HttpURLConnection，还有第三方开发包 AsyncHttpClient 等。

（1）HttpURLConnection：Android 对 HTTP 通信提供了很好的支持，通过标准的 Java 类 HttpURLConection 便可实现基于 URL 的请求及响应功能。接下来详细介绍 HttpURLConnection 的使用，代码如下：

```
1    URL url=new URL("http://www.baidu.com");
2    HttpURLConnection conn=(HttpURLConnection) url.openConnection();
3    conn.setRequestMethod("GET");
4    conn.setConnectionTimeout(5000);
5    InputStream is=conn.getInputStream();
6    conn.disconnect();
```

代码 1 行，在 URL 的构造方法中传入要访问资源的路径。

代码 2 行，调用 openConnection（）方法创建一个 HttpURLConnction 对象。

代码 3 行，设置请求方式，在使用 HttpURLConnection 访问网络时，通常会用到两种网络方式，一种是 get，一种是 post，这两种方式是在 HTTP 1.0 中定义的，用于表明 Request-URI 指定资源的不同操作方式。这两种请求方式在提交数据时也有一定的区别。

get 方式以实体的方式得到由请求 URL 所指向的资源信息，它向服务器提交的参数跟在请求 URL 后面，使用 get 方式访问网络 URL 的内容一般要小于 1KB。

使用 post 方式提交数据时，提交的数据是以键值对的形式封装在请求实体中，用户通过浏览器无法看到发送的请求数据，因此 post 方式要比 get 方式相对安全。

代码 4 行，设置超时时间。

代码 5 行，获取服务器返回的输入流。

代码 6 行，关闭 HTTP 连接。

（2）在实际开发中，使用 Android 自带的 API 与服务器进行通信比较麻烦，为了节约开发成本和时间，可以使用各种各样开源项目，下面介绍两个开源项目，AsyncHttpClient 和 SmartImageView。

AsyncHttpClient：由于访问网络是一个比较耗时的操作，在主线程中操作会出现假死或异常等情况，影响用户体验，因此谷歌规定 Android 4.0 以后访问网络的操作都必须放在子线程中。但在 Android 开发中，发送、处理 HTTP 请求十分常见，如果每次与服务器进行数据交互都需要开启一个子线程，这将是非常麻烦的，为此可以使用开源项目——AsyncHttpClient。

AsyncHttpClient 可以处理异步 HTTP 请求，并通过匿名内部类处理回调结果，HTTP 异步请求均位于非 UI 线程中，不会阻塞 UI 操作，AsyncHttpClient 通过线程池处理并发送请求，处理文件上传、下载，响应结果自动打包成 JSON 格式，使用起来非常方便，AsyncHttpClient 开源项目包含很多类，下面介绍两个常用的类。

AsyncHttpClient 类：异步客户端请求的类，提供了 get、put、post、delete、head 等请求方法。发送请求时，需要通过 AsyncHttpClient 类的实例对象访问网络。

AsyncHttpResponseHandler 类：继承自 ResponseHandlerInterface 类，访问网络后回调的接口，接收请求结果，如果访问成功则会回调 AsyncHttpResponseHandler 接口中的 OnSuccess（）方法，如果失败则会回调 OnFailure（）方法。

（3）session。

session 是浏览器和服务器会话过程中，服务器分配的一块存储空间。每次会话都会生成唯一的 sessionid，通过该 sessionid 可以确认会话的身份信息。

session 比 cookie 更安全，session 信息存储在服务器端，相对安全性更高。

服务器会给 session 一个有效期，即从该 session 的会话在有效时间内没有再被访问，其就会被设置为超时，需要重新创建会话，会话超时的时间可以通过配置文件来设置，如修改 web.xml、server.xml 文件。

（4）环境准备。

需要有 Eclipse 和 Tomcat，并在 Eclipse 中配置 Tomcat。

3.3 具体步骤

3.3.1 使用 NavicatLite 创建用户表

第 1 步：新建 SmartClass 连接。

打开 Navicat Lite，选择【连接】→【MySQL】，打开"连接"窗口，输入连接名为"SmartClass"，主机名或 IP 地址为"localhost"，域为"3306"，用户名为"root"，密码为"123"，如图 3-3 所示，单击【确定】按钮完成连接的创建。

第 2 步：创建数据库 coffeedb。

选中"SmartClass"连接，单击右键，选择【创建新数据库】，打开【创建新数据库】窗口，如图 3-4 所示，录入数据库名称为"coffeedb"，字符集选择"utf8--UTF-8 Unicode"，单击【确定】按钮完成数据库的创建。

图 3-3　新建连接 SmartClass

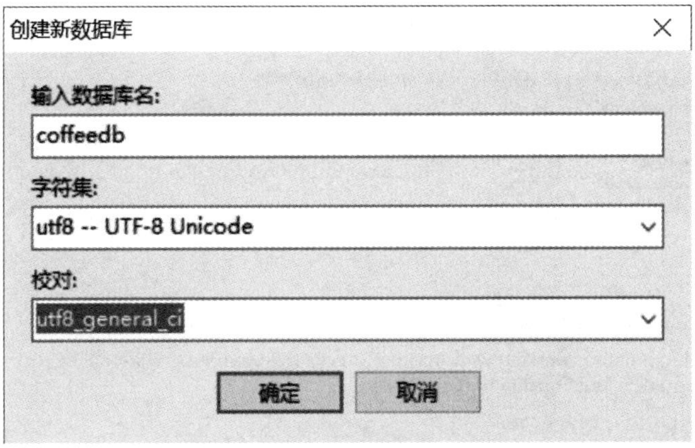

图 3-4　创建新数据库 coffeedb

第 3 步：创建用户（user）表。

在 coffeedb 数据库中创建用户（user）表，用户（user）表如表 3-1 所示。

表 3-1　用户（user）表

字段名	字段类型	是否为主键	备注
mobile	varchar(11)	是	手机号
password	varchar(6)		密码

◆ 3.3.2　创建与 Android 客户端交互的服务器端程序——与 MySQL 交互

Andoid 客户端通过 AsyncHttpCilent 向服务器端的 Servlet 发送请求并获取服务器响应，实现 Android 客户端与服务器端之间的通信。首先需要创建名称为 "SmartProductWeb" 的 Java Web 项目作为服务器端。

第 1 步：新建 Java Web 项目 SmartProductWeb。

打开 Eclipse，选择菜单【File】→【new】→【Dynamic Web Project】，打开新建项目窗口，如图 3-5 所示，在项目名称中输入 "SmartProductWeb"，单击【Next】按钮，打开如图 3-6 所示的窗口，将勾选框勾选上，单击【Finish】按钮，完成新建项目。

第 2 步：新建包。

选中 SmartProductWeb 项目的 "src" 文件夹，单击右键，选择【new】→【package】，创建名为 "db" 的包，用于存储数据库操作相关的类，同样的步骤创建名为 "entity" 的包用于存储实体类，创建名为 "servlets" 的包用于存储与客户端交互的 Servlet 相关的控制类，创建后如图 3-7 所示。

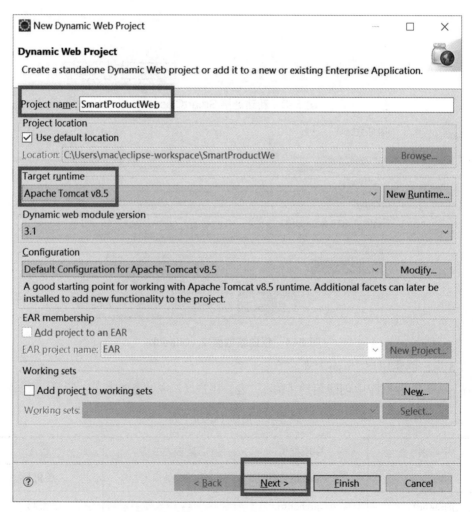

图 3-5　新建 SmartProductWeb 项目

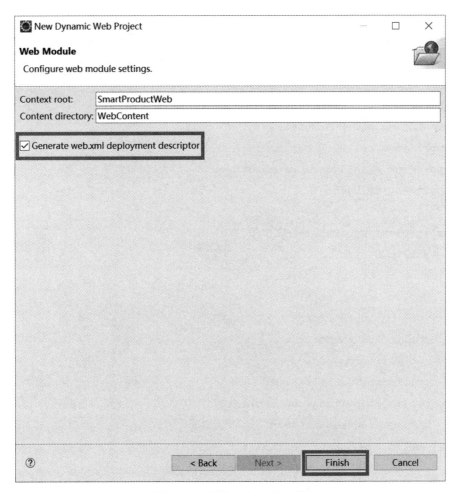

图 3-6 创建 web.xml 文件

第 3 步：创建数据库操作类（MyDatabase.java 类），用于对 MySQL 数据库进行增、删、改、查操作。

选中 "db" 包，单击右键，选择【new】→【class】，创建 MyDatabase 类，如图 3-8 所示。

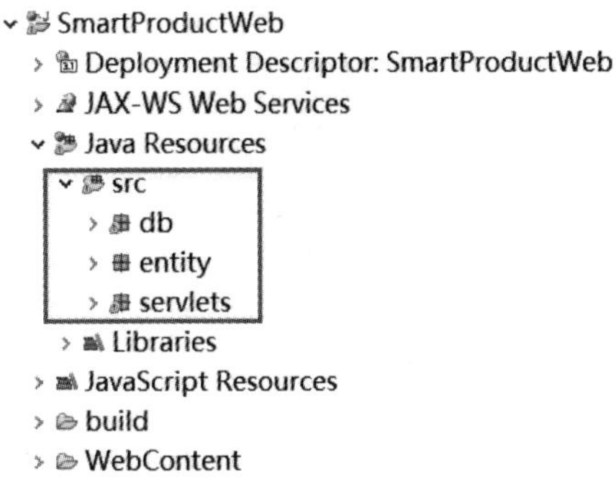

图 3-7 创建相应的包 　　　　图 3-8 创建 MyDatabase 类

使用原生的 JDBC 方式连接数据库，数据库的增、删、改、查操作借助 PreparedStatement 类和 ResultSet 类来完成。

/SmartProductWeb/src/db/MyDatabase.java 代码如下：

```java
package db;

import java.sql.Connection;
import java.sql.DriverManager;
import java.sql.PreparedStatement;
import java.sql.ResultSet;
import java.sql.SQLException;
import com.mysql.jdbc.Statement;
import entity.User;

public class MyDatabase {
    public String url;
    public String IPAddress;
    public String port;
    public String DBname;
    public String user;
    public String passwd;
    public String driver;
    public Connection conn=null;
    public PreparedStatement pst=null;
    public ResultSet rSet=null;
    // 无参数的构造函数
    public MyDatabase() {
        url="jdbc:mysql://";
        IPAddress="localhost";
        DBname="coffeedb";
        user="root";
        passwd="1234";
        driver="com.mysql.jdbc.Driver";
        try {
            Class.forName(driver);// 加载 MySQL 驱动
            url=url+IPAddress+":3306"+"/"+DBname;
            url=url+"?useUnicode=true&characterEncoding=utf-8&useSSL=false";
            conn=DriverManager.getConnection(url,user,passwd);

        } catch (ClassNotFoundException | SQLException e) {
            e.printStackTrace();
        }
    }
```

```java
39      // 有参数的构造函数
40      public MyDatabase(String driver,String hostname,String port,
        String dbname,String username,String password) {
41          url="jdbc:mysql://";
42          this.driver=driver;
43          this.IPAddress=hostname;
44          this.port=port;
45          this.DBname=dbname;
46          this.user=username;
47          this.passwd=password;
48          try {
49                  Class.forName(driver);
50                  url=url+IPAddress+":3306"+"/"+DBname;
51                  url=url+"?useUnicode=true&characterEncoding=utf-
                    8&useSSL=false";
52                  conn=DriverManager.getConnection(url,user,passwd);
53          } catch (ClassNotFoundException | SQLException e) {
54                  e.printStackTrace();
55          }
56      }
57      // 关闭数据库
58      public void closeDB() {
59          try {
60              this.conn.close();
61              this.pst.close();
62          } catch (SQLException e) {
63              e.printStackTrace();
64          }
65      }
66      /* 根据SQL语句进行查找，返回查询的结果集
67       * ResultSet 类是数据库查询的结果集
68       */
69      public ResultSet getSelectAll(String sql) {
70          try {
71              pst=conn.prepareStatement(sql);
72              rSet=pst.executeQuery();
73              return rSet;
74          } catch (SQLException e) {
75              e.printStackTrace();
76          }
```

```
77          return null;
78      }
79      // 根据SQL语句（有1个参数）进行查找，返回结果集
80      public ResultSet getSelectAll(String sql,String value) {
81          try {
82              pst=conn.prepareStatement(sql);
83              pst.setString(1,value);
84              rSet=pst.executeQuery();
85              return rSet;
86          } catch (SQLException e) {
87              e.printStackTrace();
88          }
89          return null;
90      }
91      // 根据SQL语句（带n个参数）进行查找，返回结果集
92      public ResultSet getSelectAll(String sql,String values[]) {
93          try {
94              pst=conn.prepareStatement(sql);
95              for(int i=0; i<values.length; i++) {
96                  pst.setString(i+1,values[i]);
97              }
98              rSet=pst.executeQuery();
99              return rSet;
100         } catch (SQLException e) {
101                 e.printStackTrace();
102         }
103         return null;
104     }
105     // 执行数据更新（用于删除操作）
106     public int update(String sql,String value) {
107         try {
108         if(pst!=null) pst.close();
109         pst=conn.prepareStatement(sql);
110         pst.setString(1,value);
111         return pst.executeUpdate();
112         } catch (SQLException e) {
113             e.printStackTrace();
114         }
115     return 0;
```

```java
116     }
117     // 执行数据更新（用于插入操作）
118     public int update(String sql,String values[]) {
119         try {
120             if(pst!=null) pst.close();
121             pst=conn.prepareStatement(sql);
122             for(int i=0; i<values.length; i++) {
123                 pst.setString(i+1,values[i]);
124             }
125             return pst.executeUpdate();
126         } catch (SQLException e) {
127             e.printStackTrace();
128         }
129         return 0;
130     }
131     // 执行数据更新（用于修改操作）
132     public int update(String sql,String values[],String condition) {
133         try {
134             if(pst!=null) pst.close();
135             pst=conn.prepareStatement(sql);
136             for(int i=0; i<values.length; i++) {
137                 pst.setString(i+1,values[i]);
138             }
139             pst.setString(values.length+1,condition);
140             return pst.executeUpdate();
141         } catch (SQLException e) {
142             e.printStackTrace();
143         }
144         return 0;
145     }
146     // 执行数据更新（用于修改操作，修改一个字段）
147     public int update(String sql,int value,String condition) {
148         try {
149             if(pst!=null) pst.close();
150             pst=conn.prepareStatement(sql);
151             pst.setInt(1,value);
152             pst.setString(2,condition);
153             return pst.executeUpdate();
154         } catch (SQLException e) {
```

```
155            e.printStackTrace();
156        }
157        return 0;
158    }
```

代码 23 ~ 38 行，无参数的构造函数，获取可操作的数据库对象。代码 26 ~ 28 行，是在 Navicat Lite 中创建与数据库相关信息，即数据库名称、用户名和密码，用户可以根据具体应用酌情修改。

代码 40 ~ 56 行，有参数的构造函数，获取可操作的数据库对象。

代码 58 ~ 65 行，创建函数 closeDB（），用于关闭数据库连接。

代码 69 ~ 78 行，创建函数 ResultSet getSelectAll（String sql），根据 SQL 语句进行查找，返回值为 ResultSet 类型的结果集。

代码 80 ~ 90 行，创建函数 ResultSet getSelectAll（String sql，String value），根据 SQL 语句（有 1 个参数）进行带条件查找，返回值为 ResultSet 类型的结果集。

代码 92 ~ 104 行，创建函数 ResultSet getSelectAll（String sql，String values[ ]），根据 SQL 语句（有 n 个参数）进行带条件查找，返回值为 ResultSet 类的结果集。

代码 106 ~ 116 行，创建函数 int update（String sql,String value），用于删除操作。

代码 118 ~ 130 行，创建函数 int update（String sql,String values[ ]），用于插入操作。

代码 132 ~ 145 行，创建函数 int update（String sql, String values[ ], String condition），用于修改操作。

代码 147 ~ 156 行，创建函数 int update（String sql, int value，String condition），用于修改一个字段。

◆ 3.3.3 处理用户注册的 Servlet

1. 处理用户注册的 Servlet 的作用

处理用户注册的 Servlet 只是前端控制器，它的作用有以下三个。

（1）获取客户端发送的请求参数。

（2）处理用户请求。

（3）根据处理结果生成输出。

2. 注册的接口

（1）请求地址：/SmartProductWeb/Android/RegisterServlets。

（2）请求方法：post。

（3）请求头：charset=utf-8。

（4）请求实例：mobile=15966667777&password=123。

（5）返回实例："duplicated" 或者 "success" 或者 "failed"。

第 1 步：下面创建处理用户注册的 Servlet，选中 "servlets" 包，单击右键，选择【new】→【class】，创建 RegisterServlet 类，如图 3-9 所示。

```
  SmartProductWeb
    Deployment Descriptor: SmartProductWeb
    JAX-WS Web Services
    Java Resources
      src
        db
        entity
        servlets
          RegisterServlet.java
      Libraries
    JavaScript Resources
    build
    WebContent
```

图 3-9　创建 RegisterServlet 类

/SmartProductWeb/src/servlets/RegisterServlet.java 代码如下：

```
1   package servlets;
2   
3   import java.io.IOException;
4   import java.io.PrintWriter;
5   import java.sql.ResultSet;
6   import java.sql.SQLException;
7   import javax.servlet.ServletException;
8   import javax.servlet.http.HttpServlet;
9   import javax.servlet.http.HttpServletRequest;
10  import javax.servlet.http.HttpServletResponse;
11  import db.MyDatabase;
12  
13  public class RegisterServlet extends HttpServlet {
14      private static final long serialVersionUID=1L;
15      public RegisterServlet() {
16          super();
17      }
18      protected void doPost(HttpServletRequest req,HttpServletResponse resp) throws ServletException,IOException {
19          req.setCharacterEncoding("utf-8");
20          resp.setContentType("utf-8");
21          String[] values=new String[2];
22          values[0]=req.getParameter("mobile");
23          values[1]=req.getParameter("password");
24          // 连接数据库，完成手机号查询、用户注册信息插入操作
25          MyDatabase mydatabase=new MyDatabase();
26          String insert_sql="insert into user(mobile,password) values(?,?)";
27          String query_sql="select * from user where mobile=? ";
28          ResultSet rset=mydatabase.getSelectAll(query_sql,values[0]);
29          PrintWriter out=resp.getWriter();
30          try {
```

```
31                    if(rset.next()) {
32                        out.write("duplicated");
33                    }else {
34                        int insert_result=mydatabase.update(insert_sql,
                          values);
35                        if(insert_result==1) {
36                            out.write("success");
37                        }else {
38                            out.write("failed");
39                        }
40                    }
41            } catch (SQLException e) {
42                e.printStackTrace();
43            }
44            out.flush();
45        out.close();
46    }
47    protected void doGet(HttpServletRequest request,HttpServletResponse
      response) throws ServletException,IOException {
48
49    }
50 }
```

RegisterServlet 类继承于 HttpServlet，重写了该类的 doPost（HttpServletRequest req, HttpServletResponse resp）方法，允许 Servlet 处理客户端发送过来的 post 请求，其中第 1 个参数 req，存储客户端发送来的信息；第 2 个参数 resp，存储 Servlet 要发送给客户端的信息。

代码 19～23 行，其中代码 19、20 行是指定服务器响应给客户端的编码格式和客户端请求的编码格式均为 "utf-8"，避免中文乱码；代码 22、23 行，从 req 中取出客户端发送来的信息，因为在本案例中，Andorid 客户端以 map 形式发送信息到 Servlet，因此这里调用 getParameter（String name）方法，name 为参数名，返回值为参数中存储的内容，注意这里返回值只能是 String 类或者 null。

代码 25～45 行，连接 coffeedb 数据库，查询是否存在重复数据，如果存在重复数据，则返回给服务器 "duplicated" 字符串，表示手机号有重复，注册失败；否则，将用户信息存储到数据库中的 user 表中，如果注册成功返回给服务器 "success"，表示注册成功，否则返回给服务器 "failed"，表示注册失败。

第 2 步：修改 web.xml 文件，配置 Servlet 的访问路径，客户端可以通过 "IP 地址 :8080/SmartProductWeb/Android/RegistServlet" 的地址访问到该 Servlet。

/SmartProductWeb/WebContent/WEB-INF/web.xml 代码如下：

```
1  <?xml version="1.0" encoding="UTF-8"?>
2  <web-app xmlns:xsi="http://www.w3.org/2001/XMLSchema-instance"
```

```
        xmlns="http://xmlns.jcp.org/xml/ns/javaee"
        xsi:schemaLocation="http://xmlns.jcp.org/xml/ns/javaee http://xmlns.
        jcp.org/xml/ns/javaee/web-app_3_1.xsd" id="WebApp_ID" version="3.1">
3       <display-name>SmartProductWeb</display-name>
4       <welcome-file-list>
5           <welcome-file>index.html</welcome-file>
6           <welcome-file>index.htm</welcome-file>
7           <welcome-file>index.jsp</welcome-file>
8           <welcome-file>default.html</welcome-file>
9           <welcome-file>default.htm</welcome-file>
10          <welcome-file>default.jsp</welcome-file>
11      </welcome-file-list>
12      <servlet>
13          <servlet-name>RegisterServlet</servlet-name>
14          <servlet-class>servlets.RegisterServlet</servlet-class>
15      </servlet>
16      <servlet-mapping>
17          <servlet-name>RegisterServlet</servlet-name>
18          <url-pattern>/Android/RegisterServlet</url-pattern>
19      </servlet-mapping>
20      </web-app>
```

代码 12 ~ 19 行是新添加的，其他代码是 web.xml 文件原有代码。

◆ **3.3.4 新建 LoginActivity 和 RegistActivity**

选中【java】文件夹→【com.example.coffeedemo】包→【activity】包，单击右键，选择【new】→【Activity】→【Empty Activity】，打开 Configure Activity 窗口，在【Activity Name】中输入"LoginActivity"，单击【Finish】按钮，如图 3-10 所示，即可创建登录界面。同样的步骤创建"RegistActivity"，注册界面，创建登录、注册界面后的项目目录结构如图 3-11 所示。

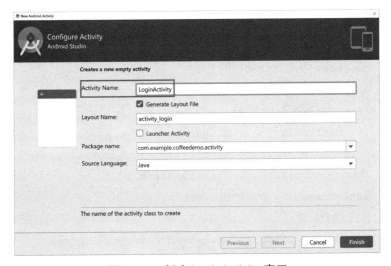

图 3-10　新建 LoginActivity 窗口

图 3-11 创建登录、注册界面后的项目目录结构

◆ 3.3.5 完成启动界面功能的逻辑实现

在启动界面,单击【登录】按钮,跳转到登录界面(见图 3-1);单击【注册】按钮,跳转到注册界面(见图 3-2),通过使用显示 Intent 完成界面之间的跳转。

java\com\example\coffeedemo\activity\MainActivity.java 代码如下:

```
1   package com.example.coffeedemo.activity;

2   import androidx.appcompat.app.AppCompatActivity;
3   import android.content.Intent;
4   import android.os.Bundle;
5   import android.view.View;
6   import android.widget.Button;
7   import com.example.coffeedemo.R;
8
9   public class MainActivity extends AppCompatActivity {
10      private Button btnLogin,btnRegist;
11      protected void onCreate(Bundle savedInstanceState) {
12          super.onCreate(savedInstanceState);
13          setContentView(R.layout.activity_main);
14          // 初始化组件
15          btnLogin=(Button)findViewById(R.id.btn_login);
16          btnRegist=(Button)findViewById(R.id.btn_regist);
```

```
17              // 跳转到登录界面
18              btnLogin.setOnClickListener(new View.OnClickListener() {
19                  public void onClick(View v) {
20                      Intent intent=new Intent(MainActivity.this,
                        LoginActivity.class);
21                      startActivity(intent);
22                  }
23              });
24              // 跳转到注册界面
25              btnRegist.setOnClickListener(new View.OnClickListener() {
26                  public void onClick(View v) {
27                      Intent intent=new Intent(MainActivity.this,
                        RegistActivity.class);
28                      startActivity(intent);
29                  }
30              });
31          }
32      }
```

关于显示 Intent 的使用，2.2.5 节已经有介绍，这里不再赘述。

3.3.6 搭建注册界面布局

布局中有带边框的效果，Android 中添加边框需要借助自定义资源，实现该效果需要两步：第 1 步，在 drawable 文件夹下，新建 XML 资源文件；第 2 步，在布局文件中通过"android:background"属性引用该资源文件即可。

第 1 步：在 drawable 文件夹下，新建 "shape_login_form.xml" 文件。

打开 "drawable" 文件夹，单击右键，选择【new】→【Drawable Resources File】，打开 New Resource File 窗口，输入 "shape_login_form"，不要输入后缀名，单击【OK】按钮完成空文件的创建。

完成样式为：矩形边框，圆角为 8dp，边框为实线，颜色为 "#f2f2f2"，内边距上、左为 1.5dp，下、右为 2dp，中心背景为白色，圆角为 10dp，内边距为 10dp。

res\drawable\shape_login_form.xml 代码如下：

```
1   <?xml version="1.0" encoding="utf-8"?>
2   <layer-list xmlns:android="http://schemas.android.com/apk/res/android">
3   <!--边-->
4       <item>
5           <shape android:shape="rectangle">
6               <padding
7                   android:bottom="2dp"
```

```
8                    android:left="1.5dp"
9                    android:right="2dp"
10                   android:top="1.5dp"/>
11          <solid android:color="#f2f2f2"/>
12          <corners android:radius="8dp"/>
13      </shape>
14  </item>
15  <!-- 中心背景 -->
16  <item>
17      <shape
18          android:shape="rectangle"
19          android:useLevel="false">
20          <solid android:color="#fff"/>
21          <corners android:radius="10dp"/>
22          <padding
23              android:bottom="10dp"
24              android:right="10dp"
25              android:top="10dp"
26              android:left="10dp"/>
27      </shape>
28  </item>
29  </layer-list>
```

第 2 步：完成界面搭建。

选择【res】→【layout】，打开"activity_regist.xml"文件，根据效果图（见图 3-2）完成界面搭建。

res\layout\activity_regist.xml 代码如下：

```
1   <?xml version="1.0" encoding="utf-8"?>
2   <RelativeLayout xmlns:android="http://schemas.android.com/apk/res/android"
3       android:layout_width="match_parent"
4       android:layout_height="match_parent">
5       <ImageView
6           android:layout_width="match_parent"
7           android:layout_height="300dp"
8           android:scaleType="fitXY"
9           android:src="@drawable/bg_login"/>
10      <LinearLayout
11          android:layout_width="match_parent"
12          android:layout_height="wrap_content"
```

```xml
13          android:orientation="vertical"
14          android:layout_alignParentBottom="true"
15          android:padding="30dp">
16      <LinearLayout
17          android:layout_width="match_parent"
18          android:layout_height="180dp"
19          android:orientation="vertical"
20          android:background="@drawable/shape_login_form"
21          android:padding="20dp">
22          <EditText
23              android:id="@+id/et_mobile"
24              android:layout_width="match_parent"
25              android:layout_height="wrap_content"
26              android:hint=" 请输入手机号 "
27              android:textSize="20sp"
28              android:layout_marginBottom="25dp"/>
29          <EditText
30              android:id="@+id/et_pwd"
31              android:layout_width="match_parent"
32              android:layout_height="wrap_content"
33              android:hint=" 请输入密码 "
34              android:textSize="20sp"/>
35      </LinearLayout>
36      <Button
37          android:id="@+id/btn_regist"
38          android:layout_width="match_parent"
39          android:layout_height="wrap_content"
40          android:text=" 注册 "
41          android:textSize="20sp"
42          android:layout_marginTop="20dp"
43          android:background="#6a1d2f"
44          android:textColor="#fff"/>
45      </LinearLayout>
46  </RelativeLayout>
```

代码 20 行，通过 "@drawable/shape_login_form" 属性，给用户输入区域加上了边框、背景等样式。

◆ **3.3.7 完成注册界面功能逻辑实现**

1. 注册功能

用户在注册界面输入手机号和密码后，将用户输入的信息发送到服务器端，服务器端对其进行验证，如果连接网络成功，则服务器端返回的结果分为以下三种。

（1）如果用户输入的手机号已经注册过，则提示"该手机号已注册"。

（2）如果用户输入的信息合法，并成功添加到数据库中，则提示"注册成功"。

（3）如果信息添加数据库失败，则提示"注册失败"。

如果连接网络失败，则提示"连接网络失败"。

2. 注册的接口

（1）请求地址：/SmartProductWeb/Android/RegisterServlet。

（2）请求方法：post。

（3）请求头：charset=utf-8。

（4）请求实例：mobile=15966667777&password=123。

（5）返回实例："duplicated"或者"success"或者"failed"。

如何实现 Android 客户端与服务器端的交互呢？方法有多种，如使用原生的 HttpClient、第三方开源项目 AsyncHttpClient，下面采用开源项目 AsyncHttpClient 实现与服务器端的交互，即实现网络编程。

第 1 步：导入开源项目 AsyncHttpClient 相关的包。

将项目目录由"Android"模式切换到"Project"模式，并将"android-async-http-1.4.8.jar"文件复制到"libs"包中，右键单击复制的 Jar 文件，选择【Add as Library】→将其导入到 Moudle 中即可。AsyncHttpClient 是第三方的开源项目，会经常被更新，使用方法可能会因为版本差异有所不同。

需要注意的是，由于 AsyncHttpClient 是对 HttpClient 的再次封装，使用某些方法时需要用到 HttpClient，因此，也需要将 HttpClient 的 Jar 文件"httpcore-4.4.4.jar"导入项目中。添加方式同 AsyncHttpClient，添加结果如图 3-12 所示。

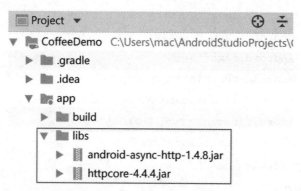

图 3-12 导入 AsyncHttpClient 需要的"jar"包后的项目目录结构

第 2 步：添加 User 实体类。

选中"entity"包，单击右键，选择【new】→【Java Class】，打开【Create New Class】窗口，输入"User"，单击【OK】按钮。创建 User 实体类后的项目目录结构如图 3-13 所示。

图 3-13　创建 User 实体类后的项目目录结构

java\com\example\coffeedemo\entity\User.java 代码如下：

```
1   package com.example.coffeedemo.entity;
2
3   public class User {
4       private String mobile;
5       private String password;
6       public void setMobile(String mobile) {
7           this.mobile=mobile;
8       }
9       public String getMobile() {
10          return mobile;
11      }
12      public void setPassword(String password) {
13          this.password=password;
14      }
15      public String getPassword() {
16          return password;
17      }
18  }
```

实体类 User.java 中设置的属性要与服务器端数据库中 user 表中的字段相对应。

第 3 步：添加 RegisterServlet 的访问地址。

打开 res/strings.xml 文件，在 <resources> 节点下添加如下一行代码：

```xml
1  <string name="registerurl">http://192.168.1.105:8080/SmartProductWeb/Android/RegisterServlet</string>
```

第 4 步：添加完成注册的功能代码。

java\com\example\coffeedemo\activity\RegistActivity.java 代码如下：

```java
1   package com.example.coffeedemo.activity;
2   
3   import androidx.appcompat.app.AppCompatActivity;
4   import android.content.Intent;
5   import android.os.Bundle;
6   import android.text.TextUtils;
7   import android.view.View;
8   import android.widget.Button;
9   import android.widget.EditText;
10  import android.widget.Toast;
11  import com.example.coffeedemo.R;
12  import com.example.coffeedemo.entity.User;
13  import com.loopj.android.http.AsyncHttpClient;
14  import com.loopj.android.http.RequestParams;
15  import com.loopj.android.http.TextHttpResponseHandler;
16  import org.apache.http.Header;
17  
18  public class RegistActivity extends AppCompatActivity{
19      private Button btnRegist;
20      private EditText etRMobile,etRPwd;
21      protected void onCreate(Bundle savedInstanceState) {
22          super.onCreate(savedInstanceState);
23          setContentView(R.layout.activity_regist);
24          // 初始化组件
25          btnRegist=(Button)findViewById(R.id.btn_regist);
26          etRMobile=(EditText)findViewById(R.id.et_mobile);
27          etRPwd=(EditText)findViewById(R.id.et_pwd);
28          // 添加按钮响应事件
29          btnRegist.setOnClickListener(new View.OnClickListener() {
30              public void onClick(View v) {
31                  String mobile=etRMobile.getText().toString().trim();
32                  String pwd=etRPwd.getText().toString().trim();
33                  if(TextUtils.isEmpty(mobile)&&TextUtils.isEmpty(pwd)){
34                      Toast.makeText(RegistActivity.this," 电话号码或密码不可以为空 ",Toast.LENGTH_SHORT).show();
```

```
35              }else{
36                  User user=new User();
37                  user.setMobile(mobile);
38                  user.setPassword(pwd);
39                  RegisterByPost(getString(R.string.registerurl),
                    user);
40              }
41          }
42      });
43  }
44  public void RegisterByPost(String url,User user){
45      RequestParams params=new RequestParams();
46      params.put("mobile",user.getMobile());
47      params.put("password",user.getPassword());
48      //实例化AsyncHttpClient对象
49      AsyncHttpClient client=new AsyncHttpClient();
50      //调用post()方法连接网络
51      client.post(url,params,new TextHttpResponseHandler() {
52          public void onFailure(int i,Header[] headers,String s,
                Throwable throwable) {
53              Toast.makeText(RegistActivity.this,"连接网络失败",
                    Toast.LENGTH_SHORT).show();
54          }
55          public void onSuccess(int i,Header[] headers,String s) {
56              if(s.equals("duplicated")){
57                  Toast.makeText(RegistActivity.this,"该手机号已注册",
                        Toast.LENGTH_SHORT).show();
58              }else if(s.equals("success")){
59                  Toast.makeText(RegistActivity.this,"注册成功",
                        Toast.LENGTH_SHORT).show();
60                  Intent intent=new Intent(RegistActivity.this,
                        LoginActivity.class);
61                  startActivity(intent);
62              }else{
63                  Toast.makeText(RegistActivity.this,"注册失败",
                        Toast.LENGTH_SHORT).show();
64              }
65          }
66      });
67  }
```

```
68    }
```

代码 29～42 行，给【注册】按钮添加点击事件监听，当【注册】按钮被点击时，获取用户的输入信息，用 Boolean TextUtils.isEmpty（String str）判断输入内容是否为空，如果为空，则提示"电话号码或密码不可以为空"；如果不为空，则将用户信息发送给服务器。

代码 44～67 行，定义了处理网络连接的内部方法 RegisterByPost（String url, User user），该方法中调用了 AsyncHttpClient 类提供的向服务器发送请求的 post（）方法。

第 5 步：修改 AndroidManifest.xml 配置文件，添加 3 个权限，即访问网络的权限、禁止明文流量的权限、使用 AsyncHttpClient 的权限。

CoffeeDemo\app\src\main\AndroidManifest.xml 代码如下：

```xml
1   <?xml version="1.0" encoding="utf-8"?>
2   <manifest xmlns:android="http://schemas.android.com/apk/res/android"
3       package="com.example.coffeedemo">
4       <uses-permission android:name="android.permission.INTERNET"/>
5       <application
6           android:allowBackup="true"
7           android:icon="@mipmap/ic_launcher"
8           android:label="@string/app_name"
9           android:roundIcon="@mipmap/ic_launcher_round"
10          android:supportsRtl="true"
11          android:theme="@style/CustomTheme"
12          android:usesCleartextTraffic="true">
13          <activity android:name=".activity.LoginActivity"></activity>
14          <activity android:name=".activity.RegistActivity"/>
15          <activity android:name=".activity.WelcomeActivity">
16              <intent-filter>
17                  <action android:name="android.intent.action.MAIN"/>
18                  <category android:name="android.intent.category.
                    LAUNCHER"/>
19              </intent-filter>
20          </activity>
21          <activity android:name=".activity.MainActivity"/>
22          <uses-library android:name="org.apache.http.legacy"
                android:required="false"/>
23      </application>
24  </manifest>
```

代码 4 行，添加访问网络的权限。

代码 12 行，如果 ComplieSdkVersion 大于 27 时，有网络安全的限制，需要请求头必须是 https 而不是 http，这时需要添加 "android:usesCleartextTraffic="true"" 属性，表示禁

用明文流量请求，配置后才可以发送 http 请求。

代码 22 行，Android 6 以后，禁用了 AsyncHttpClient，如果要使用，则需要添加如下权限：<uses-library android:name="org.apache.http.legacy" android:required="false" />。

第 6 步：运行项目。

（1）运行服务器端"SmartProductWeb"，选中该项目，单击右键，选择【Run As】→【Run on sever】，出现如图 3-14 所示内容即运行成功。

图 3-14　服务器启动成功效果图 1

（2）运行客户端 Anroid 应用程序，跳转到注册界面，注册用户信息：手机号为"15966667777"，密码为"1234"。注册成功后跳转到登录界面。

当再次进入注册界面时，输入"15966667777"，则出现"该手机号已注册"的提示，如图 3-15 所示。

图 3-15　运行注册界面效果图

课外扩展：

（1）注册界面：用户注册信息可以再丰富一些，如添加性别、昵称、短信验证码等信息。

（2）服务器端：没有采用框架，直接在 RegisterServlet 中操作数据库，RegisterServlet 中包含了实现功能的逻辑代码和数据库操作代码，代码耦合性较高，后期可以将服务器端程序进行优化，如采用 SSM 框架完成服务器端编程。

◆ 3.3.8 搭建登录界面布局

选择【res】→【layout】，打开"activity_login.xml"，根据效果图（见图 3-1）完成界面搭建。

res\layout\activity_login.xml 代码如下：

```
1   <?xml version="1.0" encoding="utf-8"?>
2   <RelativeLayout xmlns:android="http://schemas.android.com/apk/res/android"
3       android:layout_width="match_parent"
4       android:layout_height="match_parent">
5       <ImageView
6           android:layout_width="match_parent"
7           android:layout_height="300dp"
8           android:scaleType="fitXY"
9           android:src="@drawable/bg_login"/>
10      <LinearLayout
11          android:layout_width="match_parent"
12          android:layout_height="wrap_content"
13          android:orientation="vertical"
14          android:layout_alignParentBottom="true"
15          android:padding="30dp">
16          <LinearLayout
17              android:layout_width="match_parent"
18              android:layout_height="180dp"
19              android:orientation="vertical"
20              android:background="@drawable/shape_login_form"
21              android:padding="20dp">
22              <EditText
```

```
23              android:id="@+id/et_mobile"
24              android:layout_width="match_parent"
25              android:layout_height="wrap_content"
26              android:hint=" 请输入手机号 "
27              android:textSize="20sp"
28              android:layout_marginBottom="20dp"/>
29          <EditText
30              android:id="@+id/et_pwd"
31              android:layout_width="match_parent"
32              android:layout_height="wrap_content"
33              android:hint=" 请输入密码 "
34              android:textSize="20sp"/>
35      </LinearLayout>
36      <Button
37          android:id="@+id/btn_login"
38          android:layout_width="match_parent"
39          android:layout_height="wrap_content"
40          android:text=" 登录 "
41          android:textSize="20sp"
42          android:layout_marginTop="25dp"
43          android:background="#6a1d2f"
44          android:textColor="#fff"/>
45  </LinearLayout>
46 </RelativeLayout>
```

代码 20 行,通过 "@drawable/shape_login_form" 属性,给用户输入区域加上了边框、背景等样式。

◆ 3.3.9 处理用户登录的 Servlet

1. 处理用户登录的 Servlet 的作用

处理用户登录的 Servlet 只是前端控制器,它的作用有以下三个。

(1)获取客户端发送的请求参数。

(2)处理用户请求。

(3)根据处理结果生成 JSON 输出。

2. 登录的接口

(1)请求地址:/SmartProductWeb/Android/LoginServlets。

(2)请求方法:post。

（3）请求头：charset=utf-8。

（4）请求实例：mobile=15966667777&password=123。

（5）返回实例：登录成功，则返回{"mobile"："15966667777"，"msessionid"：动态生成的sessionid}；登录失败，则返回{"mobile"：""}。

第1步：添加GSON依赖包。

将GSON依赖包gson-2.7.jar文件复制到"SmartProductWeb"项目的lib目录下，如图3-16所示。

第2步：创建处理用户登录的Servlet，选中"servlets"包，单击右键，选择【new】→【class】，创建LoginServlet类，如图3-17所示。

图3-16　添加gson-2.7.jar包　　　图3-17　创建LoginServlet类

第3步：配置LoginServlet的访问路径。

将LoginServlet添加到web.xml中的<web-app>节点下，客户端可以通过"IP地址:8080/SmartProductWeb/Android/LoginServlet"的地址访问到该Servlet，代码如下：

```
1    <servlet>
2        <servlet-name>LoginServlet</servlet-name>
3        <servlet-class>servlets.LoginServlet</servlet-class>
4    </servlet>
5    <servlet-mapping>
6        <servlet-name>LoginServlet</servlet-name>
7          <url-pattern>/Android/LoginServlet</url-pattern>
8    </servlet-mapping>
```

第4步：修改LoginServlet类，完成该功能。

接收客户端传来的手机号和密码，连接数据库进行验证，验证登录成功后，将用户的唯一标识手机号存储在session中，并将用户手机号和sessionid以JSON格式返回给Android客户端。

/SmartProductWeb/src/servlets/LoginServlet.java代码如下：

```java
1   package servlets;
2   
3   import java.io.IOException;
4   import java.sql.ResultSet;
5   import java.sql.SQLException;
6   
7   import javax.servlet.ServletException;
8   import javax.servlet.http.HttpServlet;
9   import javax.servlet.http.HttpServletRequest;
10  import javax.servlet.http.HttpServletResponse;
11  import javax.servlet.http.HttpSession;
12  
13  import com.google.gson.JsonObject;
14  
15  import db.MyDatabase;
16  
17  public class LoginServlet extends HttpServlet {
18  
19      private static final long serialVersionUID=1L;
20  
21      public LoginServlet() {
22          super();
23      }
24  
25      protected void doPost(HttpServletRequest req,HttpServletResponse resp) throws ServletException,IOException {
26          req.setCharacterEncoding("utf-8");
27          resp.setContentType("application/json,charset=utf-8");
28  
29          String[] values=new String[2];
30          values[0]=req.getParameter("mobile");
31          values[1]=req.getParameter("password");
32  
33          MyDatabase mydatabase=new MyDatabase();
34          String query_sql="select * from user where mobile=? and password=?";
35          ResultSet rset=mydatabase.getSelectAll(query_sql,values);
36          try {
37                  // 登录成功
38                  if(rset.next()) {
39                      req.getSession(true).setAttribute("mobile",values[0]);
40                      // 把验证的 mobile 封装成 JsonObject
41                      JsonObject jsonObj=new JsonObject();
```

```
42                    jsonObj.addProperty("mobile",values[0]);
43                    jsonObj.addProperty("msessionid",req.getSession().
                      getId());
44                    resp.getWriter().println(jsonObj);
45                }else {
46                    JsonObject jsonObj=new JsonObject();
47                    jsonObj.addProperty("mobile","");
48                    resp.getWriter().println(jsonObj);
49                }
50            }catch (SQLException e) {
51                e.printStackTrace();
52            }
53            mydatabase.closeDB();
54        }
55    }
```

LoginServet 类继承于 HttpServlet，重写了该类的 doPost（HttpServletRequest req, HttpServletResponse resp）方法，允许 Servlet 处理客户端发送过来的 post 请求，其中：第 1 个参数 req 存储客户端发送来的信息；第 2 个参数 resp 存储 Servlet 要发送给客户端的 JSON 数据。

客户端向服务器发送 /SmartProductWeb/Android/LoginServlet 请求，如果登录成功将得到如图 3-18 所示的 JSON 响应数据。

```
{"mobile":"15966667777","msessionid":"68E1B039BC5A6A4061AC900E818176F5"}
```

图 3-18　登录成功的 JSON 响应

图 3-18 所示的字符串就是典型的 JSON 格式字符串，Android 客户端程序只要调用 JSONObject 的 getString（String key, String defvalue）即可取到其值。

注意，其中的"msessionid"的值是 Android 客户端与服务器端会话时，服务器端生成的唯一标识，不是固定不变的。

代码 26 行，req 对象调用 setCharacterEncoding（"utf-8"）方法指定客户端发送过来的消息的编码格式为"utf-8"。

代码 27 行，resp 对象调用 setContentType（"application/json,charset=utf-8"）方法设置服务器端发送给客户端的信息为 JSON 数据，编码格式为 utf-8。

setCharacterEncoding（String charset）方法的参数为编码格式，如"utf-8""GBK"等格式。setContentType（String type）方法的参数可以是编码格式，也可以是包含编码格式的内容类型，根据实际应用场景选择合适的内容类型或者编码格式，如 setContentType（"utf-8"）或者 setContentType（"application/json, charset=utf-8"）。

代码 29 ~ 31 行，从 req 中取出客户端发送过来的手机号和密码，存储在数组 values 中。

代码 33 ~ 35 行，查询手机号和密码是否正确，并将查询得到的记录存储在 ResultSet

类型的变量 rset 中。

代码 38～49 行，如果 rset.next（）为 true，则表示查询的结果集不为空，登录成功；否则，登录失败。如果登录成功，代码 39 行，req 对象调用 getSession（）方法可以获得 HttpSession 类型的对象，并将用户的手机号存入 session 对象中，Android 客户端其他界面可以从该 session 中取出用户的手机号，实现登录状态的保持。代码 41～44 行，调用 jsonObj.addProperty（"mobile"，values[0]）和 jsonObj.addProperty（"msessionid"，req.getSession（）.getId（））生成 JSON 对象，并调用 resp.getWriter（）.println（jsonObj）方法将 JSON 对象返回给客户端。代码 46～48 行，如果登录失败，则调用 jsonObj.addProperty（"mobile"，""）方法，将 {"mobile"，""} 这个 JSON 对象返回给客户端。

◆ 3.3.10　新建 HomeActivity

选中"java"文件夹→【com.example.coffeedemo】包→【activity】包，单击右键，选择【new】→【Activity】→【Empty Activity】，打开 Configure Activity 窗口，在【Activity Name】输入框中输入"HomeActivity"，单击【Finish】按钮，如图 3-19 所示，即可完成首界面的创建。首界面布局和功能逻辑代码将在第 4 章完成，本节只是创建空界面，作为登录成功后跳转的界面。

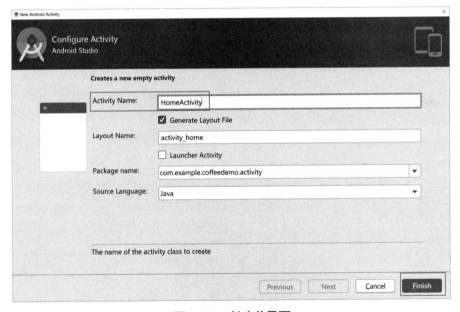

图 3-19　创建首界面

◆ 3.3.11　完成登录界面功能逻辑实现

1. 功能描述

登录时，用户输入注册的手机号和密码，单击【登录】按钮，将用户输入的信息发送到服务器端进行验证：如果验证成功，则跳转到 App 的首界面；如果验证失败，则提示用户"手机号或密码错误，请重新输入"。

与之前不同的是，本次请求时返回的数据为 JSON 格式的数据。

2. 登录的接口

（1）请求地址：/SmartProductWeb/Android/LoginServlet。

（2）请求方法：post。

（3）请求头："application/json,charset=GBK"。

（4）请求实例：mobile=15966667777&password=123。

（5）响应实例：登录成功，则返回 { "mobile"："15966667777"，"msessionid"：动态生成的 sessionid}；登录失败，则返回 { "mobile"：""}。

第 1 步：添加 LoginServlet 的访问地址。

打开 res/strings.xml 文件，在 <resources> 节点下添加如下代码：

```
1   <string name="loginurl">http://192.168.1.105:8080/SmartProductWeb/Android/LoginServlet</string>
```

第 2 步：完成登录界面功能的逻辑实现。

java\com\example\coffeedemo\activity\LoginActivity.java 代码如下：

```
1   package com.example.coffeedemo.activity;
2
3   import androidx.appcompat.app.AppCompatActivity;
4   import android.content.Intent;
5   import android.os.Bundle;
6   import android.text.TextUtils;
7   import android.view.View;
8   import android.widget.Button;
9   import android.widget.EditText;
10  import android.widget.Toast;
11  import com.example.coffeedemo.R;
12  import com.example.coffeedemo.entity.User;
13  import com.loopj.android.http.AsyncHttpClient;
14  import com.loopj.android.http.JsonHttpResponseHandler;
15  import com.loopj.android.http.RequestParams;
16  import com.loopj.android.http.TextHttpResponseHandler;
17  import org.apache.http.Header;
18  import org.json.JSONException;
19  import org.json.JSONObject;
20  public class LoginActivity extends AppCompatActivity {
21      private Button btnLogin;
22      private EditText etRMobile,etRPwd;
23      protected void onCreate(Bundle savedInstanceState) {
24          super.onCreate(savedInstanceState);
25          setContentView(R.layout.activity_login);
26          // 初始化组件
27          btnLogin=(Button)findViewById(R.id.btn_login);
```

```java
28        etRMobile=(EditText)findViewById(R.id.et_mobile);
29        etRPwd=(EditText)findViewById(R.id.et_pwd);
30        // 添加登录按钮点击事件
31        btnLogin.setOnClickListener(new View.OnClickListener() {
32            public void onClick(View v) {
33                String mobile=tRMobile.getText().toString().trim();
34                String pwd=etRPwd.getText().toString().trim();
35                if(TextUtils.isEmpty(mobile)||TextUtils.isEmpty(pwd)){
36                    Toast.makeText(LoginActivity.this," 电话号码或密码不可以为空 ",Toast.LENGTH_SHORT).show();
37                }else{
38                    User user=new User();
39                    user.setMobile(mobile);
40                    user.setPassword(pwd);
41                    LoginByPost(getString(R.string.loginurl),user);
42                }
43            }
44        });
45    }
46    public void LoginByPost(String url,User user){
47        RequestParams params=new RequestParams();
48        params.put("mobile",user.getMobile());
49        params.put("password",user.getPassword());
50        // 实例化 AsyncHttpClient 对象
51        AsyncHttpClient client=new AsyncHttpClient();
52        // 调用 post() 方法连接网络
53        client.post(url,params,new JsonHttpResponseHandler(){
54            public void onFailure(int statusCode,Header[] headers, Throwable throwable,JSONObject errorResponse) {
55                super.onFailure(statusCode,headers,throwable,errorResponse);
56                Toast.makeText(LoginActivity.this," 连接网络失败 ",Toast.LENGTH_SHORT).show();
57            }
58            public void onSuccess(int statusCode,Header[] headers, JSONObject response) {
59                super.onSuccess(statusCode,headers,response);
60                try {
61                    if(response.getString("mobile").isEmpty()){
62                        Toast.makeText(LoginActivity.this," 手机号或密码错误，请重新输入 ",Toast.LENGTH_SHORT).show();
63                    }else{
64                        SharedPreferences sp=getSharedPreferences("login",MO DE_PRIVATE);
```

```
65                    SharedPreferences.Editor editor=sp.edit();
66                    editor.putString("msessionid",response.
                      getString("msessionid"));
67                    editor.commit();
68                    Intent intent=new Intent(LoginActivity.this,
                      HomeActivity.class);
69                    startActivity(intent);
70                    finish();
71                   }
72                  } catch (JSONException e) {
73                      e.printStackTrace();
74              }
75          }
76      });
77  }
78 }
```

代码 35 ～ 42 行，判断用户输入的内容是否为空，如果为空，则提示用户"电话号码或密码不可以为空"；如果不为空，则调用自定义方法 LoginByPost（String url, User user）连接服务器的 LoginServlet，判断用户输入是否正确。

代码 46 ～ 77 行，是连接服务器的 LoginServlet 的方法 LoginByPost（String url, User user）。关于 AsyncHttpClient 类这里不再赘述，之前已经介绍过。代码 64 ～ 70 行，判断服务器返回的手机号是否为空：如果为空，表示用户输入的手机号或密码错误，登录失败；如果不为空，表示登录成功，调用方法 response.getString（"msessionid"）获得服务器为本次会话生成的 sessionid，并保存到共享参数文件 login.xml 文件中，如图 3-20 和图 3-21 所示（图 3-20 和图 3-21 需要在执行完第 2 步，运行项目才能查看到），然后跳转到首界面 HomeActivity，并关闭当前 LoginActivity 界面。

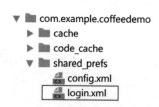

图 3-20　DeviceExplore 查看共享参数文件

```
login.xml ×
1  <?xml version='1.0' encoding='utf-8' standalone='yes' ?>
2  <map>
3      <string name="msessionid">68E1B039BC5A6A4061AC900E818176F5</string>
4  </map>
```

图 3-21　login.xml 文件内容

第 3 步：运行项目。

（1）运行服务器端"SmartProductWeb"，选中该项目，单击右键，选择【Run As】→【Run

on sever】，出现如图 3-22 所示内容即运行成功。

图 3-22　服务器启动成功效果图 2

（2）运行客户端 Android 应用程序，跳转到登录界面，有如下三种场景。

①输入错误的手机号或密码，如输入手机号"15966667777"，密码"admin"，单击【登录】按钮，登录失败，如图 3-23 所示。

②没有输入手机号或密码，如仅输入手机号，没有输入密码，单击【登录】按钮，登录失败，如图 3-24 所示。

③输入正确的手机号和密码，如输入手机号"15966667777"，密码"123"，单击【登录】按钮，登录成功，跳转到空的首界面，并关闭登录界面。

图 3-23　登录失败（密码不对）　　图 3-24　登录失败（没有输入密码，提示输入密码）

课外扩展：

为了实现登录状态的保持，有的应用程序采用将登录成功的用户编号等敏感信息存储在共享参数文件中。但实际上，这是不够安全的，如果应用程序登录状态不在服务器端进行控制，而只在客户端控制是十分脆弱的。

对恶意用户来说，他可以用多种方法获得存储在手机上的敏感信息，这是不够安全的。

第 4 章

首界面、咖啡列表和详情界面

学习目标

通过咖啡列表和详情界面的实战学习,学生能够分析程序的基本功能,利用所学知识完成界面的搭建和功能实现,具体目标如下。

(1) 了解 Fragment 的应用场合和创建方法,以及 Activity 和 Fragment 之间的调用,理解它们的声明周期关系。

(2) 理解 <RadioButton> 组件,运用所学知识实现界面底部的导航,单击单选按钮界面也会随之变化。

(3) 掌握使用 ViewPager 实现滑动特效。

(4) 理解 <GridView> 和 Intent 数据传递的应用场景。

(5) 掌握使用 GSON 解析 JSON 数据的方法。

(6) 掌握 Android 客户端与服务器端通信的方法。

4.1 任务描述

首界面由一个 Fragment 和多个单选按钮组成的，咖啡列表界面包含了轮播特效和咖啡商品展示，点击商品可以查看商品详情。

图 4-1　首界面、咖啡列表界面

图 4-2　咖啡详情界面

4.2 相关知识

◆ 4.2.1　Fragment 布局

Android 3.0 引入了 Fragment 的概念，主要是支持大屏幕上更为动态和灵活的 UI 设计，如平板电脑。通过将 Activity 的布局分割成若干个 Fragment，可以在运行时编辑 Activity 的呈现，并且这些变化会被保存在由 Activity 管理的后台栈里面。

一个 Activity 可以包含多个 Fragment，一个 Fragment 也可以在多个 Activity 中使用。

我们知道 Activity 的生命周期有 5 种状态，分别是启动状态、运行状态、暂停状态、停止状态和销毁状态，Fragment 的生命周期也有这几种状态。

因为 Fragment 是被嵌入到 Activity 中使用的，因此它的生命周期状态直接受其所属 Activity 的生命周期状态影响。当在 Activity 中创建 Fragment 时，Fragment 处于启动状态；当 Activity 被暂停时，所有在该 Activity 中的 Fragment 也被暂停；当 Activity 被销毁时，所有在该 Activity 中的 Fragment 也被销毁。

◆ 4.2.2　RadioButton 组件

<RadioButton> 为单选按钮，它常与 <RadioGroup> 配合使用，提供两个或多个按钮互斥的选项集，在 <RadioGroup> 中可以利用"android:orientation"属性控制 <RadioButton>

的排列方向。

在 XML 布局文件中，RadioGroup 和 RadioButton 配合使用，具体代码如下：

```
1  <RadioGroup
2      android:layout_width=" 设置宽度 "
3      android:layout_height=" 设置高度 "
4      android:orientation=" 设置方向 ">
5  <RadioButton
6      android:layout_width=" 设置宽度 "
7      android:layout_height=" 设置高度 "
8      android:text=" 设置显示的文本信息 " />
9  <RadioButton
10     android:layout_width=" 设置宽度 "
11     android:layout_height=" 设置高度 "
12     android:text=" 设置显示的文本信息 " />
13 </RadioGroup>
```

（1）<RadioGroup> 继承自 LinearLayout，可以使用 android:orientation 属性控制 <RadioButton> 的排列方向。

（2）android:text：设置单选按钮显示的文本信息。

◆ 4.2.3 ViewPage 组件

ViewPager 是 Android 3.0 后引入的一个 UI 组件，视图滑动切换工具，通过手势滑动可以完成 View 的切换。

ViewPager 的创建步骤是，首先在布局文件中添加 ViewPager 组件，然后创建 PagerAdapter，最后设置 onPagerChanger 监听器。

◆ 4.2.4 SmartImageView 组件

SmartImageView 的出现主要是方便从网络上加载图片，它继承自 Android 自带的 ImageView 组件，另外它还附加了功能。例如，支持根据 URL 地址加载图片，支持加载通讯录中的图片，支持异步加载图片，支持图片缓存等。

SmartImageView 在使用之前，同样需要将 SmartImageView 的 Jar 文件导入项目中，使用步骤如下。

（1）在布局文件中添加 SmartImageView 控件，代码如下：

```
1  <com.loopj.android.image.SmartImageView
2      android:id="@+id/sivIcon"
3      android:layout_width="match_parent"
4      android:layout_height="match_parent">
5  </com.loopj.android.image.SmartImageView>
```

（2）在 Java 文件中使用 SmartImageView 控件，代码如下：

```
1   SmartImageView siv=(SmartImageView)findViewById(R.id.sivIcon);
2   url="http://localhost:8080/img/img.jpg";
3   siv.setImageUrl(url,R.mipmap.icon_launcher,R.mipmap.ic_launcher);
```

4.3 具体步骤

◆ 4.3.1 Fragment+RadioButon 实现底部导航栏

在首界面中实现底部导航功能，首界面就是第 3 章中新建的 HomeActivity。

第 1 步：搭建首界面的布局，包括 FrameLayout 和 RadioButton 等。

res\layout\activity_home.xml 代码如下：

```
1   <?xml version="1.0" encoding="utf-8"?>
2   <LinearLayout xmlns:android="http://schemas.android.com/apk/res/android"
3       android:layout_width="match_parent"
4       android:layout_height="match_parent"
5       android:orientation="vertical">
6   <FrameLayout
7       android:id="@+id/fragment_main"
8       android:layout_width="match_parent"
9       android:layout_height="match_parent"
10      android:layout_weight="1"/>
11  <RadioGroup
12      android:id="@+id/main_nav"
13      android:layout_width="match_parent"
14      android:layout_height="wrap_content"
15      android:orientation="horizontal">
16      <RadioButton
17          android:id="@+id/home"
18          android:layout_width="wrap_content"
19          android:layout_height="wrap_content"
20          android:drawableTop="@drawable/icon_home_gray"
21          android:checked="true"
22          style="@style/RadioGroupButtonMainNavStyle"/>
23      <RadioButton
24          android:id="@+id/chat"
25          android:layout_width="wrap_content"
26          android:layout_height="wrap_content"
27          android:drawableTop="@drawable/icon_chat_gray"
28          style="@style/RadioGroupButtonMainNavStyle"/>
29      <RadioButton
```

```
30                    android:id="@+id/wode"
31                    android:layout_width="wrap_content"
32                    android:layout_height="wrap_content"
33                    android:drawableTop="@drawable/icon_wo_gray"
34                    style="@style/RadioGroupButtonMainNavStyle"/>
35            </RadioGroup>
36    </LinearLayout>
```

使用单帧布局 <FrameLayout> 和 <RadioButton> 配合实现底部导航栏，使用 <RadioGroup> 可以实现单选按钮的互斥选中效果。

代码22行，用到了自定义样式，需要在样式表中自定义名为"RadioGroupButtonMainNavStyle"的样式，实现一次定义多次使用，减少代码的冗余，实现方法如下。

打开 res/values/styles.xml 文件，在 <resources> 节点下添加如下代码：

```
1    <style name="RadioGroupButtonMainNavStyle">
2        <item name="android:button">@null</item>
3        <item name="android:gravity">center</item>
4        <item name="android:layout_width">0dp</item>
5        <item name="android:layout_height">40dp</item>
6        <item name="android:layout_weight">1</item>
7        <item name="android:textSize">0sp</item>
8    </style>
```

第2步：创建3个Fragment，并为每个Fragment创建一个布局。

（1）选中"com.example.coffeedemo"文件夹，单击右键，选择【new】→【新建package】，新建"fragment"包。

（2）选中"fragment"包，单击右键，选择【new】→【Fragment】→【Fragment（Blank）】，新建3个Frgment，分别是"HomeFragment""CartFragment"和"WodeFragment"。新建完成后项目目录结构如图4-3和图4-4所示。

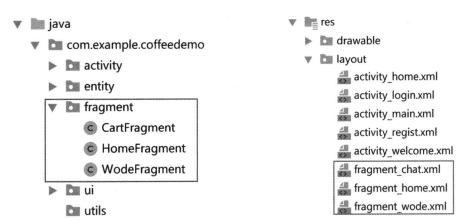

图4-3　3个Fragment应对的Java文件　　图4-4　创建3个Fragment对应的布局文件

第3步：点击单选按钮实现Fragment的切换，通过setOnCheckedChangeListener（）

方法为RadioGroup对象注册监听器，在onCheckedChanged（）方法中，使用switch来判断哪个单选按钮被选中，进而实现对应Fragment的加载。

java\com\example\coffeedemoactivity\HomeActivity.java 代码如下：

```java
package com.example.coffeedemo.activity;

import androidx.appcompat.app.AppCompatActivity;
import androidx.fragment.app.FragmentManager;
import androidx.fragment.app.FragmentTransaction;
import android.os.Bundle;
import android.widget.RadioGroup;
import com.example.coffeedemo.R;
import com.example.coffeedemo.fragment.CartFragment;
import com.example.coffeedemo.fragment.HomeFragment;
import com.example.coffeedemo.fragment.WodeFragment;

public class HomeActivity extends AppCompatActivity {
    private RadioGroup main_nav;
    private FragmentManager fragmentManager;
    private FragmentTransaction fragmentTransaction;

    protected void onCreate(Bundle savedInstanceState) {
        super.onCreate(savedInstanceState);
        setContentView(R.layout.activity_home);
        // 实现底部导航
        buildMainNav();
    }
    public void buildMainNav() {
        main_nav=(RadioGroup) findViewById(R.id.main_nav);
        // 设置默认导航
        fragmentManager=getSupportFragmentManager();
        fragmentTransaction=fragmentManager.beginTransaction();
        HomeFragment homeFragment=new HomeFragment();
        fragmentTransaction.replace(R.id.fragment_main,homeFragment);
        fragmentTransaction.addToBackStack(null);
        fragmentTransaction.commit();
        main_nav.setOnCheckedChangeListener(new RadioGroup.OnCheckedChangeListener() {
            public void onCheckedChanged(RadioGroup group,int checkedId) {
                fragmentManager=getSupportFragmentManager();
```

```
36              fragmentTransaction=fragmentManager.beginTransaction();
37              switch (checkedId) {
38                  case R.id.home:
39                      HomeFragment homeFragment=new HomeFragment();
40                      fragmentTransaction.replace(R.id.fragment_main,
                        homeFragment);
41                      break;
42                  case R.id.chat:
43                      CartFragment cartfragment=new CartFragment();
44                      fragmentTransaction.replace(R.id.fragment_main,
                        cartfragment);
45                      break;
46                  case R.id.wode:
47                      WodeFragment wodefragment=new WodeFragment();
48                      fragmentTransaction.replace(R.id.fragment_main,
                        wodefragment);
49                      break;
50              }
51              fragmentTransaction.addToBackStack(null);
52              fragmentTransaction.commit();
53          }
54      });
55  }
56 }
```

第 4 步：运行 CoffeeDemo 项目即可看到效果。

课外扩展：

还可以增加界面切换效果，主要涉及的知识点为 PageAdapter、ViewPager、Fragment 模块，掌握使用 ViewPager 实现滑动特效。

◆ **4.3.2 搭建咖啡列表界面布局**

咖啡列表放在首界面中，HomeFragment 对应首界面，修改 fragment_home.xml 文件以完成首界面布局，使用 ViewPager 实现轮播效果，使用 <GridView> 实现咖啡列表。

res/layout/fragment_home.xml 代码如下：

```
1  <?xml version="1.0" encoding="utf-8"?>
2  <LinearLayout xmlns:android="http://schemas.android.com/apk/res/android"
```

```xml
3       android:layout_width="match_parent"
4       android:layout_height="match_parent"
5       android:orientation="vertical"
6       android:padding="10dp">
7       <!-- 轮播图 -->
8       <RelativeLayout
9           android:id="@+id/slider_contianer"
10          android:layout_width="match_parent"
11          android:layout_height="150dp"
12          android:orientation="vertical">
13          <androidx.viewpager.widget.ViewPager
14              android:id="@+id/pager"
15              android:layout_width="match_parent"
16              android:layout_height="150dp"/>
17          <!-- 位置点父容器 -->
18          <LinearLayout
19              android:id="@+id/lyDot"
20              android:orientation="horizontal"
21              android:layout_width="wrap_content"
22              android:layout_height="wrap_content"
23              android:layout_marginBottom="10dp"
24              android:layout_alignParentBottom="true"
25              android:layout_centerHorizontal="true">
26          </LinearLayout>
27      </RelativeLayout>
28      <!-- 商品展示 -->
29      <TextView
30          android:layout_width="wrap_content"
31          android:layout_height="wrap_content"
32          android:text=" 潮品 "
33          android:textSize="18sp"
34          android:textStyle="bold"
35          android:layout_marginBottom="10dp"/>
36      <GridView
37          android:id="@+id/gvGoods"
38          android:layout_width="match_parent"
39          android:layout_height="wrap_content"
40          android:numColumns="2"
41          android:verticalSpacing="10dp"
```

```
42            android:horizontalSpacing="10dp"/>
43      </LinearLayout>
```

4.3.3 新建咖啡详情界面

选中"java"文件夹→【com.example.coffeedemo】包→【activity】包，单击右键，选择【new】→【Activity】→【Empty Activity】，打开 Configure Activity 窗口，在【Activity Name】输入框中输入"GoodsDetailActivity"，单击【Finish】按钮，如图 4-5 所示，即可完成咖啡详情界面的创建。

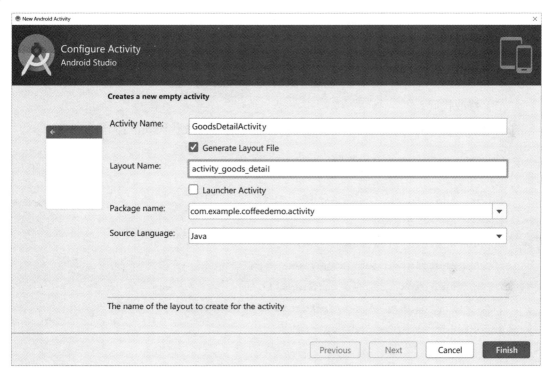

图 4-5　创建咖啡详情界面

接下来需要完成轮播效果的功能实现和咖啡商品列表功能的实现。

4.3.4 实现轮播特效

第 1 步：在界面中添加〈ViewPager〉节点，可以参见 4.3.2 中代码 13～16 行。

第 2 步：创建 PageAdapter 适配器。

（1）选中"com.example.coffeedemo"文件夹，单击右键，选择【new】→【新建 package】，新建"ui"包。

（2）选中"ui"包，单击右键，选择【new】→【Java Class】，打开如图 4-6 所示窗口，输入类名"SliderPageAdapter"，在继承的类【Superclass】中输入类的关键字"PageAdapter"，在下拉提示框中选择"androidx.viewpager.widget.PagerAdapter"，然后单击【OK】按钮。

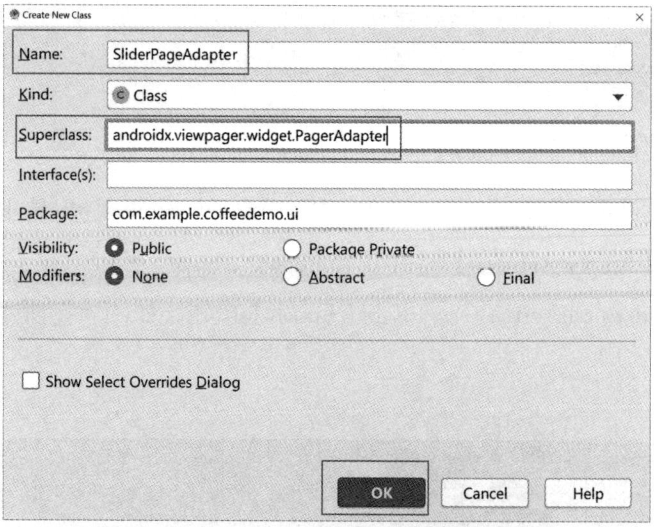

图 4-6 创建"SliderPageAdapter"类

（3）修改代码。

java\com\example\coffeedemo\ui\SliderPageAdapter.java 修改代码如下：

```
1   package com.example.coffeedemo.ui;
2
3   import android.content.Context;
4   import android.view.View;
5   import android.view.ViewGroup;
6   import android.widget.ImageView;
7   import androidx.annotation.NonNull;
8   import androidx.viewpager.widget.PagerAdapter;
9   import com.example.coffeedemo.R;
10
11  public class SliderPageAdapter extends PagerAdapter{
12      private Context mContext;
13      private int[] mData;
14      /*
15       * 构造函数
16       * 初始化上下文和数据
17       * @param context
18       * @param list
19       */
20      public SliderPageAdapter(Context context,int[] data){
21          mContext=context;
22          mData=data;
23      }
24      /*
```

```java
25      * 返回要滑动的view个数
26      */
27     @Override
28     public int getCount() {
29         return mData.length;
30     }
31
32     /*
33      * 1.将当前视图添加到container中
34      * 2.返回当前view
35      */
36     @NonNull
37     @Override
38     public Object instantiateItem(@NonNull ViewGroup container,int position) {
39         View view=View.inflate(mContext,R.layout.slider_item_img,null);
40         ImageView iv=view.findViewById(R.id.img);
41         iv.setScaleType(ImageView.ScaleType.FIT_XY);
42         iv.setImageResource(mData[position]);
43         container.addView(view);
44         return view;
45     }
46     /*
47      * 从当前container中删除指定位置position的view
48      */
49
50     @Override
51     public void destroyItem(@NonNull ViewGroup container,int position,@NonNull Object object) {
52         //super.destroyItem(container,position,object);
53         container.removeView((View)object);
54     }
55
56     /*
57      * 确定视图是否与特定键对象关联
58      */
59     @Override
60     public boolean isViewFromObject(@NonNull View view,@NonNull Object object) {
61         return view==object;
```

```
62      }
63  }
```

第 3 步：设置 onPageChange 监听器，实现轮播特效。

java\com\example\coffeedemo\fragment\HomeFragment.java 代码如下：

```
1   package com.example.coffeedemo.fragment;
2
3   import android.content.Intent;
4   import android.os.Bundle;
5   import android.os.Handler;
6   import android.os.Message;
7   import android.view.LayoutInflater;
8   import android.view.View;
9   import android.widget.ImageView;
10  import android.widget.LinearLayout;
11
12  import androidx.fragment.app.Fragment;
13  import androidx.viewpager.widget.ViewPager;
14
15  import com.example.coffeedemo.R;
16  import com.example.coffeedemo.ui.SliderPageAdapter;
17
18  public class HomeFragment extends Fragment {
19      private ViewPager viewPager;
20      private int[] imgData={R.drawable.slider_img_1,R.drawable.slider_img_2,R.drawable.slider_img_3,R.drawable.slider_img_4};
21
22      private Handler handler;
23      private int currentPosition=0;// 设置 ViewPager 当前位置
24      /* 当前 activity 是否存活，当为 false 时，结束 ViewPager 轮播线程 */
25      private boolean actIsAlive=true;
26      private LinearLayout lyDot;
27
28      public View onCreateView(LayoutInflater inflater,ViewGroup container, Bundle savedInstanceState) {
29          View view=inflater.inflate(R.layout.fragment_home,container, false);
30          viewPager=(ViewPager)view.findViewById(R.id.pager);
31          lyDot=(LinearLayout)view.findViewById(R.id.lyDot);
32          // 使用 ViewPager 实现轮播
```

```java
33            initSliderCont();
34            initHandler();
35            autoViewPager();
36            initDots(view);
37            return view;
38        }
39        private void initSliderCont(){
40            SliderPageAdapter adapter=new SliderPageAdapter(getContext(),imgData);
41            viewPager.setAdapter(adapter);
42            viewPager.setOnPageChangeListener(new ViewPager.OnPageChangeListener() {
43                @Override
44                public void onPageScrolled(int position,float positionOffset,int positionOffsetPixels) {
45                }
46                public void onPageSelected(int position) {
47                    currentPosition=position;
48                    for(int i=0;i<imgData.length;i++){
49                        if(i==currentPosition){
50                            ImageView imageView=(ImageView)lyDot.getChildAt(i);
51                            imageView.setImageResource(R.drawable.bg_dot_active);
52                        }else{
53                            ImageView imageView=(ImageView)lyDot.getChildAt(i);
54                            imageView.setImageResource(R.drawable.bg_dot);
55                        }
56                    }
57                }
58                public void onPageScrollStateChanged(int state) {
59                }
60            });
61        }
62        private void initHandler() {
63            handler=new Handler() {
64                @Override
65                public void handleMessage(Message msg) {
66                    super.handleMessage(msg);
```

```java
67                if (msg.what==1) {
68                    if (currentPosition==imgData.length-1){
69                        currentPosition=0 ;
70                        viewPager.setCurrentItem(0,false);
71                    }else{
72                        currentPosition++;
73                        viewPager.setCurrentItem(currentPosition,true);
74                    }
75                }
76            }
77        };
78    }
79    /**
80     * ViewPager 自动播放
81     */
82    private void autoViewPager() {
83        new Thread() {
84            @Override
85            public void run() {
86                super.run();
87                while(actIsAlive) {
88                    try {
89                        sleep(3000);
90                        handler.sendEmptyMessage(1);
91                    } catch (InterruptedException e) {
92                        e.printStackTrace();
93                    }
94                }
95            }
96        }.start();
97    }
98    /**
99     * 动态创建轮播图位置点显示
100    */
101   private void initDots(View view) {
102       // 动态添加轮播图位置点，默认第 0 个位置为当前轮播图的颜色
103       for(int i=0; i<imgData.length; i++) {
104           ImageView imageView=new ImageView(view.getContext());
105
106           if(i==0) {
```

```
107                    imageView.setImageResource(R.drawable.bg_dot);
108                }else{
109                    imageView.setImageResource(R.drawable.bg_dot_active);
110                }
111                lyDot.addView(imageView);
112            }
113        }
114    }
```

接下来实现咖啡列表功能，第 1 步需要创建 Goods 实体类，第 2 步定义 <GridView> 布局中的每一项布局，第 3 步为 <GridView> 中咖啡商品数据做准备，第 4 步定义适配器，第 5 步给 <GridView> 中的每一项添加点击事件监听，点击时跳转到商品详情界面。

◆ 4.3.5 创建 Goods 实体类

在"entity"包下定义一个 Good 类，如图 4-7 所示，在 Goods 类中设置商品 id、商品名称、商品价格等属性。考虑跳转到咖啡详情界面时要显示咖啡的详细信息，即要将咖啡对象传到咖啡详情界面，所以要在 Goods 类中实现 Serializable 接口。

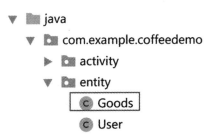

图 4-7 在"entity"包下定义 Goods 类后的项目目录结构

java\com\example\coffeedemo\entity\Goods.java 代码如下：

```
1   package com.example.coffeedemo.entity;
2
3   import java.io.Serializable;
4
5   public class Goods implements Serializable {
6       private String goodsId;
7       private String goodsImg;
8       private String title;
9       private float price;
10      private int soldNums;
11      private String des;
12
13      public String getGoodsId(){
14          return goodsId;
```

```
15      }
16      public String getImg(){
17          return goodsImg;
18      }
19      public String getTitle(){
20          return title;
21      }
22      public float getPrice(){
23          return price;
24      }
25      public int getSold_nums(){
26          return soldNums;
27      }
28      public void setImg(int img){
29          this.goodsImg=goodsImg;
30      }
31      public String getDes(){
32          return des;
33      }
34
35      public void setGoodsId(String goodsId){
36          this.goodsId=goodsId;
37      }
38      public void setTitle(String title){
39          this.title=title;
40      }
41      public void setPrice(float price){
42          this.price=price;
43      }
44      public void setSold_nums(int sold_nums){
45          this.soldNums=sold_nums;
46      }
47      public void setDes(String des){
48          this.des=des;
49      }
50  }
```

◆ 4.3.6 定义 <GridView> 布局中的每一项布局

选中【layout】，单击右键，选择【new】→【Layout resource file】，打开如图 4-8 所

示界面,输入名称"goods_gv_item",然后单击【OK】按钮即可完成。

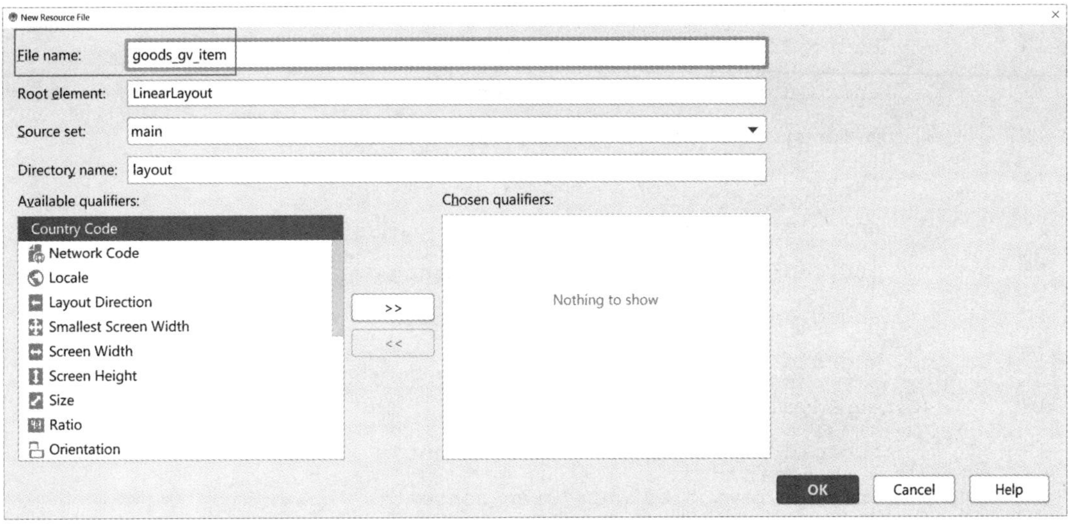

图 4-8 创建 <GridView> 每一项布局 goods_gv_item.xml

该布局定义咖啡列表中每一项的布局,根据效果图(见图 4-9),这里加载图片的组件使用第三方控件 SmartImageView,可以方便加载网络图片。在使用 SmartImageView 之前,需要将 SmartImageView 的"jar"包导入项目中。

将 CoffeeDemo 项目切换到"Project"视图模式下,将"android-smart-image-view-1.0.0.jar"包拷贝到"libs"文件夹下,右键单击复制的 Jar 文件,选择【Add As Library】,导入项目即可。

图 4-9 每一项的布局效果图

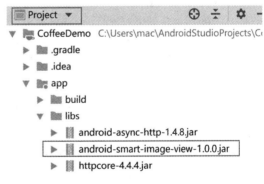

图 4-10 导入 SmartImageView 的"jar"包

res/goods_gv_item.xml 代码如下:

```
1  <?xml version="1.0" encoding="utf-8"?>
2  <LinearLayout xmlns:android="http://schemas.android.com/apk/res/android"
3      android:orientation="vertical"
4      android:layout_width="wrap_content"
5      android:layout_height="wrap_content">
6      <com.loopj.android.image.SmartImageView
7          android:id="@+id/sivIcon"
```

```
8            android:layout_width="match_parent"
9            android:layout_height="70dp"
10           android:layout_marginBottom="5dp"
11           android:scaleType="centerCrop">
12      </com.loopj.android.image.SmartImageView>
13      <TextView
14           android:id="@+id/gv_goods_title"
15           android:layout_width="wrap_content"
16           android:layout_height="wrap_content"
17           android:text=" 我是标题 "/>
18      <LinearLayout
19           android:layout_width="match_parent"
20           android:layout_height="wrap_content"
21           android:orientation="horizontal"
22           android:padding="5dp">
23           <TextView
24               android:id="@+id/gv_goods_price"
25               android:layout_width="wrap_content"
26               android:layout_height="wrap_content"
27               android:text=" 价格 "/>
28           <TextView
29               android:id="@+id/gv_goods_solded_nums"
30               android:layout_width="match_parent"
31               android:layout_height="wrap_content"
32               android:text=" 销量 "
33               android:gravity="right"
34               android:textSize="12sp"
35               android:textColor="#d4143c"/>
36      </LinearLayout>
37  </LinearLayout>
```

4.3.7 咖啡商品数据准备

咖啡商品信息来源于数据库中，客户端连接服务器端获取数据库中存储的咖啡商品信息，将其展示在客户端。

第 1 步：在数据库中创建 goods 表，存储商品基本信息。

打开 NavicatLite，找到 coffeedb 数据库，创建商品信息（goods）表（见表 4-1），并录入基础数据。

第 4 章
首界面、咖啡列表和详情界面

表 4-1 商品信息（goods）表

字段名	字段类型	是否为主键	备注
goodsId	varchar(255)	是	商品编号
goodsImg	varchar(255)		商品图片地址
title	text		商品名字
price	float		商品价格
soldNums	int(11)		商品已售数量
des	text		商品描述

客户端连接服务器端，服务器端从数据库中将商品信息查询出来，以 JSON 格式返回给客户端，返回 JSON 数据如下：

```
1  [
2    {
3      "goodsID":"001",
4      "price":35,
5      "title":"卡布奇诺",
6      "des":"经典奶咖，奶泡与咖啡交融，绵密醇香，轻盈如雪。",
7      "soldNums":10,
8      "goodsImg":"http://192.168.1.105:8080/img/img_american.jpg"
9    },
10   {
11     "goodsID":"002",
12     "price":35,
13     "title":"焦糖玛奇朵",
14     "des":"经焦糖风味奶咖，上层注入丰富奶泡，层次感分明。",
15     "soldNums":10,
16     "goodsImg":"http://192.168.1.105:8080/img/img_greentea.jpg"
17   },
18   {
19     "goodsID":"003",
20     "price":35,
21     "title":"加浓美式",
22     "des":"比标准美式更多一份 Espresso，口感更加香醇浓厚，回味持久，清醒加倍。",
23     "soldNums":10,
24     "goodsImg":"http://192.168.1.105:8080/img/img_latte.jpg"
25   },
26   {
27     "goodsID":"004",
28     "price":28,
```

```
29        "title":" 榛果拿铁 ","des":" 榛果爱好者的选择，香甜榛果风味与咖啡牛奶融合，
          诠释一种新鲜风味。",
30        "soldNums":10,
31        "goodsImg":"http://192.168.1.105:8080/img/img_macchiato.jpg"
32      }
33    ]
```

由此可以发现，goodsImg 中存储的是图片的地址，如代码 8 行提供了图片地址，客户端的 SmartImageView 控件将会加载地址为"http://192.168.1.105:8080/img/img_american.jpg"的图片。

"http://192.168.1.105:8080/img/img_american.jpg"这个地址表示什么意思呢？表示访问本机安装在 Tomcat → webapps → ROOT 目录下的 img 文件夹中的名为"img_american.jpg"的图片。

需要 2 步，首先将图片放到本地的 Tomcat 相应目录下，找到本机安装的 Tomcat 目录，找到 webapps 下的 ROOT 目录，如找到"C:\software\apache-tomcat-8.5.40\webapps\ROOT\"，新建"img"文件夹，将用到的图片拷贝到此目录下。

然后修改此处的 IP 地址为本机的 IP 地址。

◆ 4.3.8　定义咖啡商品用到的适配器

第 1 步：新建 GoodsAdapter 适配器。

选中"ui"包，单击右键，选择【new】→【Java Class】，打开如图 4-11 所示窗口，输入类名"GoodsAdapter"，在继承的类【Superclass】中输入类的关键字"BaseAdapter"，在下拉提示框中选择"android.widget.BaseAdapter"，然后单击【OK】按钮。

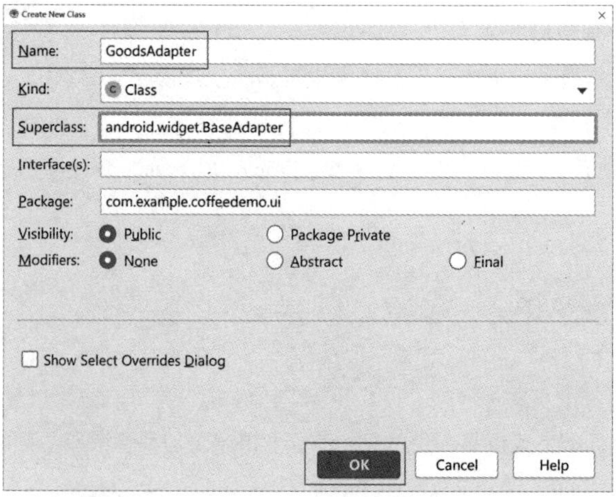

图 4-11　创建"GoodsAdapter"类

第 2 步：完善代码。

java\com\example\coffeedemo\ui\GoodsAdapter.java 代码如下：

```
1   package com.example.coffeedemo.ui;
2
```

```java
3   import android.content.Context;
4   import android.view.View;
5   import android.view.ViewGroup;
6   import android.widget.BaseAdapter;
7   import android.widget.TextView;
8   import com.example.coffeedemo.entity.Goods;
9   import com.example.coffeedemo.R;
10  import com.example.coffeedemo.entity.Goods;
11  import com.loopj.android.image.SmartImageView;
12  import java.util.List;
13
14  public class GoodsAdapter extends BaseAdapter {
15      private Context myContext;
16      private List<Goods> mylist;
17
18      public GoodsAdapter(Context context,List<Goods> productsList) {
19          myContext=context;
20          mylist=productsList;
21      }
22      public int getCount() {
23          return mylist.size();
24      }
25      public Object getItem(int position) {
26          return position;
27      }
28      public long getItemId(int position) {
29          return position;
30      }
31      public View getView(int position,View convertView,ViewGroup parent) {
32          View view;
33          /*view重用 */
34          if(convertView==null){
35              view=View.inflate(myContext,R.layout.goods_gv_item,null);
36          }else{
37              view=convertView;
38          }
39          SmartImageView siv=(SmartImageView)view.findViewById(R.id.sivIcon);
40          //SmartImageView加载指定图片路径
41          siv.setImageUrl(mylist.get(position).getImg(),R.mipmap.ic_launcher,R.mipmap.ic_launcher);
42          TextView title=view.findViewById(R.id.gv_goods_title);
```

```
43          title.setText(mylist.get(position).getTitle());
44          TextView price=view.findViewById(R.id.gv_goods_price);
45          price.setText(mylist.get(position).getPrice()+" 元 ");
46          TextView sold_nums=view.findViewById(R.id.gv_goods_solded_
            nums);
47          sold_nums.setText(mylist.get(position).getSold_nums()+"");
48          return view;
49      }
50  }
```

◆ 4.3.9 处理商品信息的 Servlet

1. 处理商品信息的 Servlet 的作用

处理商品信息的 Servlet 只是前端控制器，它的作用有以下三个。

（1）获取客户端发送的请求参数。

（2）处理用户请求。

（3）根据处理结果，将商品信息生成 JSON 类型数据，并输出给客户端，数据格式参见 4.3.7。

2. 商品信息的接口

（1）请求地址：/SmartProductWeb/Android/GoodsServlet。

（2）请求方法：post。

（3）请求头：charset=utf-8。

（4）请求实例：空。

（5）返回实例：商品信息以 JSON 格式返回。

第 1 步：创建处理商品信息的 Servlet，选中"servlets"包，单击右键，选择【new】→【class】，创建 GoodsServlet 类，如图 4-12 所示。

```
v 📂 SmartProductWeb
  > 🗎 Deployment Descriptor: SmartProductWeb
  > 🔷 JAX-WS Web Services
  v 📂 Java Resources
    v 📂 src
      > 📦 db
      > 📦 entity
      v 📦 servlets
        > 📄 GoodsServlet.java
        > 📄 LoginServlet.java
        > 📄 RegisterServlet.java
    > 📂 Libraries
  > 📂 JavaScript Resources
  > 📂 build
  > 📂 WebContent
```

图 4-12 创建 GoodsServlet 类

第 2 步：配置 GoodsServlet 的访问路径。

将 GoodsServlet 添加到 web.xml 中的 <web-app> 节点下，客户端可以通过"IP 地址 :8080/SmartProductWeb/Android/GoodsServlet"的地址访问到该 Servlet，添加内容如下：

```xml
<servlet>
    <servlet-name>GoodsServlet</servlet-name>
    <servlet-class>servlets.GoodsServlet</servlet-class>
</servlet>
<servlet-mapping>
    <servlet-name>GoodsServlet</servlet-name>
    <url-pattern>/Android/GoodsServlet</url-pattern>
</servlet-mapping>
```

第 3 步：完成 GoodsServlet 的功能。

查询 goods 表中的所有数据，以 JSON 格式返回给 Android 客户端。

/SmartProductWeb/src/servlets/GoodsServlet.java 代码如下：

```java
package servlets;

import java.io.IOException;
import java.io.PrintWriter;
import java.sql.ResultSet;
import java.sql.SQLException;
import javax.servlet.ServletException;
import javax.servlet.http.HttpServlet;
import javax.servlet.http.HttpServletRequest;
import javax.servlet.http.HttpServletResponse;
import com.google.gson.JsonArray;
import com.google.gson.JsonObject;
import db.MyDatabase;

public class GoodsServlet extends HttpServlet {
    private static final long serialVersionUID=-2083579833918441543L;
    public GoodsServlet() {}
    protected void doPost(HttpServletRequest req,HttpServletResponse resp) throws ServletException,IOException {
        req.setCharacterEncoding("utf-8");
        resp.setCharacterEncoding("utf-8");
        resp.setContentType("text/html,charset=utf-8");
        MyDatabase mydatabase=new MyDatabase();
        String query_sql="select * from goods";
        ResultSet rSet=mydatabase.getSelectAll(query_sql);
        JsonArray mjsonarray=new JsonArray();
```

```
26          try {
27              while(rSet.next()) {
28                  JsonObject jsonObj=new JsonObject();
29                  jsonObj.addProperty("goodsId",rSet.getString(1));
30                  jsonObj.addProperty("goodsImg",rSet.getString(2));
31                  jsonObj.addProperty("title",rSet.getString(3));
32                  jsonObj.addProperty("price",rSet.getFloat(4));
33                  jsonObj.addProperty("soldNums",rSet.getInt(5));
34                  jsonObj.addProperty("des",rSet.getString(6));
35                  mjsonarray.add(jsonObj);
36              }
37          } catch (SQLException e) {
38              e.printStackTrace();
39          }
40          mydatabase.closeDB();
41          resp.getWriter().out.print(mjsonarray.toString());
42      }
43  }
```

客户端向服务器发送 /SmartProductWeb/Android/GoodsServlet 请求，如果成功将看到如图 4-13 所示的 JSON 响应数据。

```
[{"goodsId":"1","goodsImg":"http://192.168.1.105:8080/img/img_american.jpg","title":"卡布奇诺","price":35.0,"soldNums":10,"des":"经典奶咖，奶泡与咖啡交融，绵密醇香，轻盈如雪。"},
{"goodsId":"2","goodsImg":"http://192.168.1.105:8080/img/img_greentea.jpg","title":"焦糖玛奇朵","price":35.0,"soldNums":10,"des":"经焦糖风味奶咖，上层注入丰富奶泡，层次感分明。"},
{"goodsId":"3","goodsImg":"http://192.168.1.105:8080/img/img_latte.jpg","title":"加浓美式","price":35.0,"soldNums":5,"des":"比标准美式更多一份Espresso，口感更加香醇浓厚，回味持久，清醒加倍。"},
{"goodsId":"4","goodsImg":"http://192.168.1.105:8080/img/img_macchiato.jpg","title":"榛果拿铁","price":35.0,"soldNums":20,"des":"榛果拿铁\",\"des\":\"榛果爱好者的选择，香甜榛果风味与咖啡牛奶融合，诠释一种新鲜风味。"}]
```

图 4-13　查询商品的 JSON 响应

图 4-13 所示的字符串就是 JSON 格式字符串，JSONArray 包含 JSONObject，Android 客户端接收到这样的数据后，调用 GSON 将其解析成 Goods 实体类对象存储在数组列表中。

代码 28 ~ 35 行，将数据库中查询出的所有商品集数据转换为 JsonArray 类型，JsonArray 中每一个元素都是 JsonObject 类型，存储每一种商品。这里采用 addProperty（String key，String value）方法一一添加属性，当商品属性较多时，这显然会很不方便，如果能借助实体类则更方便，读者可自行完善。

代码 41 行，将 JsonArray 类型的数据 mjsonarray 转换为字符串类型，并输出给客户端。

◆ **4.3.10　完成咖啡列表界面功能逻辑实现**

客户端调用 GoodsServlet 返回所有商品信息，商品信息以 JSON 格式存储，然后界面使用 GSON 解析服务器端返回的 JSON 数据，并展示到界面上。最后给每一个商品添加点击事件监听，点击时跳转到商品详情界面。

第 1 步：在 HomeFragment 类中添加 GridView 类型变量、Goods 类型变量声明，定义 List<Goods> 类型变量存储所有商品信息，代码如下：

```
1    private GridView goodsGridView;
2    private Goods product;
3    private List<Goods> goodsList=new ArrayList<Goods>();
```

第 2 步：客户端添加 GSON 依赖包。

商品信息以 JSON 格式存储，客户端使用 GSON 解析 JSON 文件，需要将 GSON 依赖包导入项目。在工程目录中，打开 build.gradle（Module:app）文件，如图 4-14 所示，找到 dependencies，添加如下一行代码：

```
1    implementation 'com.google.code.gson:gson:2.8.5'
```

在窗口右上角，点击窗口右上角出现的"Sync Now"，下载依赖包完成同步即可，如图 4-15 所示。

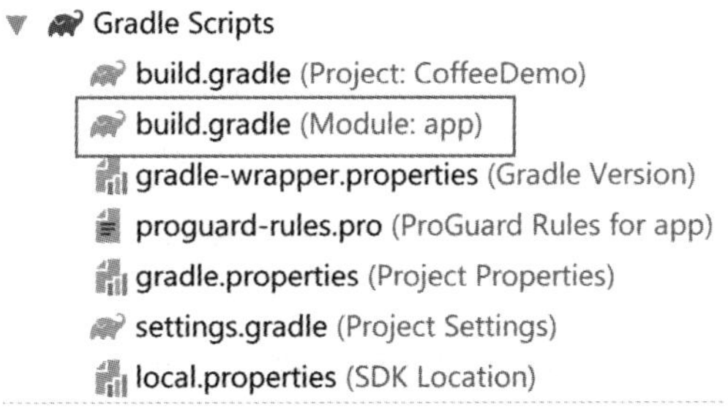

图 4-14　builde.gradle(Module:app) 位置

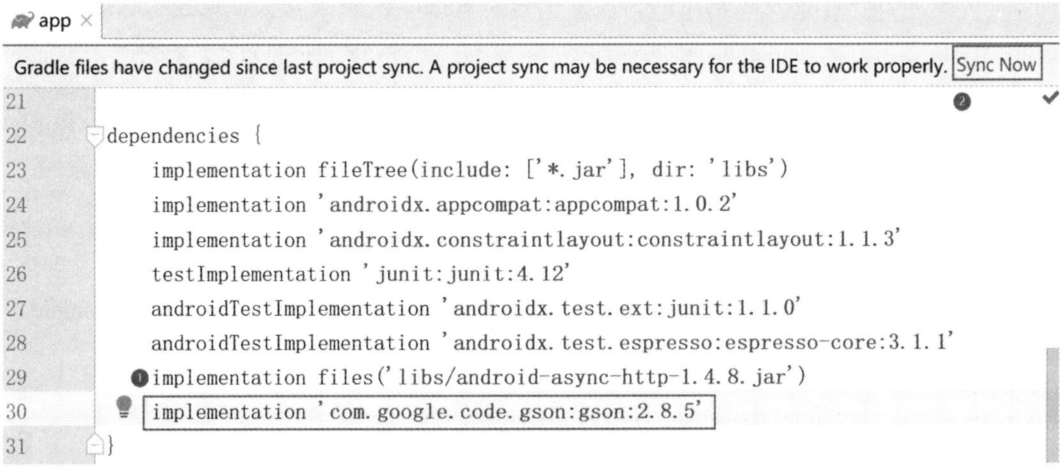

图 4-15　添加 GSON 依赖

第 3 步：添加 GoodsServlet 的访问地址。

打开 res/strings.xml 文件，在 <resources> 节点下添加如下一行代码：

```
1    <string name="goodsurl">http://192.168.1.105:8080/SmartProductWeb/
     Android/GoodsServlet</string>
```

第 4 步：添加商品详情界面。

因为在 HomeFragment 中要实现点击商品就跳转到商品详情界面，这里先创建空的商品详情界面，商品详情界面内容将在 4.3.11 节中完成。

选中"java"文件夹→【com.example.coffeedemo】包→【activity】包，单击右键，选择【new】→【Activity】→【Empty Activity】，打开 Configure Activity 窗口，在【Activity Name】输入框中输入"GoodsDetailActivity"，单击【Finish】按钮，如图 4-16 所示，即可完成商品详情界面的创建。

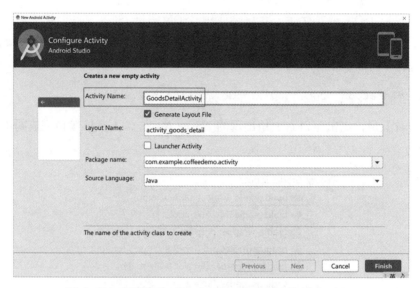

图 4-16　创建名为"GoodsDetailActivity"的 Activity

第 5 步：在 onCreateView（ ）方法中，添加 <GridView> 布局容器的定义，并实现商品展示，向 /android/GoodsServlet 发送请求时，需要携带 sessionid 来表明该状态是登录状态，商品展示调用自定义方法 void getGoodsFromServer（String url,String msessionid）来完成，添加的代码如下：

```
1    goodsGridView=(GridView)view.findViewById(R.id.gvGoods);
2    /* 实现商品展示 */
3    getGoodsFromServer(getString(R.string.goodsurl),getSessionid());
```

在 HomeFragment 类中，添加 void getGoodsFromServer（String url，String msessionid）方法的定义，代码如下：

```
1    public void getGoodsFromServer(String url,String msessionid){
2        // 实例化 AsyncHttpClient 对象
3        AsyncHttpClient client=new AsyncHttpClient();
4        client.addHeader("cookie","JSESSIONID="+msessionid);
5        // 调用 post() 方法连接网络
6        client.post(url,new TextHttpResponseHandler() {
7            public void onFailure(int i,Header[] headers,String s,
```

```java
                    Throwable throwable) {
8               Toast.makeText(getContext(),"连接网络失败",Toast.
                LENGTH_SHORT).show();
9           }
10          public void onSuccess(int i,Header[] headers,String s) {
11              goodsList=parseJson(s);
12              if(goodsList==null){
13                  Toast.makeText(getContext(),"解析失败",Toast.
                    LENGTH_SHORT).show();
14              }else{
15                  //创建适配器，加载ListView数据
16                  GoodsAdapter adapter=new GoodsAdapter(getContext(),
                    goodsList);
17                  goodsGridView.setAdapter(adapter);
18                  goodsGridView.setOnItemClickListener(new
                    AdapterView.OnItemClickListener() {
19                      public void onItemClick(AdapterView<?>
                        parent,View view,int position,long id) {
20                          Intent intent=new Intent(getContext(),
                            GoodsDetailActivity.class);
21                          intent.putExtra("goods",goodsList.
                            get(position));
22                          startActivity(intent);
23                      }
24                  });
25              }
26          }
27      });
28  }
29  public List<Goods> parseJson(String json){
30      //使用Gson库解析JSON数据
31      Gson gson=new Gson();
32      //创建一个TypeToken的匿名子类对象，并调用对象的getType()方法
33      Type listType=new TypeToken<List<Goods>>(){}.getType();
34      //把获取到的信息集合存到newsInfos中
35      List<Goods> goodsList=gson.fromJson(json,listType);
36      return goodsList;
37  }
38  public String getSessionid(){
39      SharedPreferences sp=getSharedPreferences("login",MODE_PRIVATE);
40          String msessionid=sp.getString("msessionid","");
```

```
41          return msessionid;
42    }
```

依然使用 AsyncHttpClient 类提供的 post（）方法实现客户端与服务器端的交互，本次调用的是 post（String url, new TextHttpResponseHandler（）{}）方法，第 1 个参数为要访问的 GoodsServlet 地址，即 http://192.168.1.105:8080/SmartProductWeb/Android/GoodsServlet，第 2 个参数表示服务器端的 Servlet 将返回给客户端文本类型的数据。

代码 4 行，调用 client.addHeader（"cookie"，"JSESSIONID=" + msessionid）方法将登录时生成的 sessionid 设置到请求中。

代码 10～27 行，如果连接 GoodsServlet 成功，则解析、展示商品信息，并添加每一项商品的点击事件，实现点击商品就跳转到商品详情界面查看商品详情。

代码 11 行，调用自定义方法 List<Goods> parseJson（String json）解析 JSON 格式的数据，该方法将根据 Goods 实体类中的属性解析 JSON 数据，goodsList 变量中存储的每一个元素都是一个咖啡商品对象。

代码 18～24 行，调用 GridView 类对象 goodsGridView 的 setOnItemClickListener（）方法，给每一项添加点击事件监听，点击时就跳转到商品详情界面。代码 21 行，调用 intent.putExtra（"goods"，goodsList.get（position））将"商品对象"传递给商品详情界面，实现了对象类型数据的传递。

第 6 步：运行项目。

（1）运行服务器端"SmartProductWeb"，选中该项目，单击右键，选择【Run As】→【Run on sever】，出现如图 4-17 所示内容即运行成功。

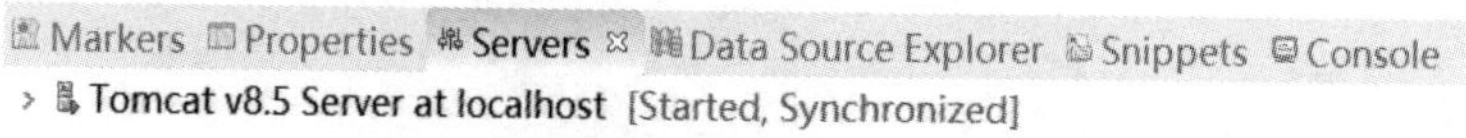

图 4-17　服务器启动成功效果图

（2）运行客户端 Android 应用程序，可以得到如图 4-1 所示效果图。

课外扩展：

（1）商品列表界面可以引入延迟加载，下拉刷新。

因为商品信息可能很多，有时 <GridView> 需要加载很多的数据，如果数据量足够大，就可能会让这个加载数据变得非常慢，即使是使用新线程来加载数据，也会等待较长的时间，于是下拉刷新/上拉刷新的功能就出现了，每次可能只显示 10 条或者 20 条数据，然后当用户下拉/上拉到最后一条/第一条数据，就再次加载 10 条或者 20 条数据，这样可以大大提升 App 的反应速度。

（2）商品信息远远不止本案例中这些属性，如口味、类别等，课后可根据实际情况进行适当扩展。

◆ **4.3.11 搭建咖啡商品详情界面布局**

咖啡详情界面展示咖啡详情，根据效果图（见图 4-2）完成界面布局，这里需要使用第三方组件 SmartImageView 来展示商品大图。

res/layout/activity_goods_detail.xml 代码如下：

```
1   <?xml version="1.0" encoding="utf-8"?>
2   <LinearLayout xmlns:android="http://schemas.android.com/apk/res/android"
3       android:layout_width="match_parent"
4       android:layout_height="match_parent"
5       android:orientation="vertical"
6       android:padding="10dp">
7       <com.loopj.android.image.SmartImageView
8           android:id="@+id/sivImg"
9           android:layout_width="match_parent"
10          android:layout_height="94dp"
11          android:layout_marginBottom="5dp"
12          android:scaleType="centerCrop"></com.loopj.android.image.SmartImageView>
13      <TextView
14          android:id="@+id/goodsDes"
15          android:layout_width="match_parent"
16          android:layout_height="wrap_content"
17          android:textSize="18sp"
18          android:text=" 我是描述 "
19          android:layout_marginTop="20dp"
20          android:layout_weight="1"/>
21      <Button
22          android:id="@+id/addToCart"
23          android:layout_width="match_parent"
24          android:layout_height="wrap_content"
25          android:text=" 加入购物车 "
26          android:textColor="#fff"
27          android:background="#8B4513"
28          android:textSize="18sp"
29          android:textStyle="bold"
30          android:layout_marginTop="40dp"
31          android:layout_marginBottom="20dp"/>
32  </LinearLayout>
```

◆ **4.3.12 完成咖啡详情界面功能逻辑实现**

在咖啡列表中，点击咖啡商品就跳转到对应的咖啡详情界面，咖啡详情界面从 Intent 中取出传递过来的咖啡对象，展示传递过来的咖啡信息。

java\com\example\coffeedemo\activity\GoodsDetailActivity.java 代码如下：

```java
package com.example.coffeedemo.activity;

import androidx.appcompat.app.AppCompatActivity;
import android.content.Intent;
import android.os.Bundle;
import android.view.View;
import android.widget.Button;
import android.widget.TextView;
import com.example.coffeedemo.R;
import com.example.coffeedemo.entity.Goods;
import com.loopj.android.image.SmartImageView;

public class GoodsDetailActivity extends AppCompatActivity {
    private TextView goodsDes;
    private SmartImageView siv;
    private Button btnAdd;

    protected void onCreate(Bundle savedInstanceState) {
        super.onCreate(savedInstanceState);
        setContentView(R.layout.activity_goods_detail);
        // 初始化组件
        goodsDes=(TextView)findViewById(R.id.goodsDes);
        siv=(SmartImageView)findViewById(R.id.sivImg);
        btnAdd=(Button)findViewById(R.id.addToCart);
        // 显示商品详情
        Intent intent=getIntent();
        Goods goods=(Goods)intent.getSerializableExtra("goods");
        siv.setImageUrl(goods.getImg());
        goodsDes.setText(goods.getDes());
        // 添加商品到购物车
        btnAdd.setOnClickListener(new View.OnClickListener() {
            @Override
            public void onClick(View v) {
                // 在第 5 章补充此处代码
            }
```

```
36              });
37          }
38    }
```

代码 26、27 行，调用 getIntent（）方法获得列表界面传递来的 intent 类的对象 intent，intent.getSerializableExtra（"goods"）可以获得传递来的 Goods 对象。

代码 28、29 行，设置详情图片和咖啡详情介绍。

代码 31 ~ 36 行，调用 setOnClickListener（）方法，给界面底部的【加入购物车】按钮添加点击事件的监听，点击时执行的功能将在第 5 章中介绍。

运行项目，运行客户端 Android 应用程序，点击咖啡列表中的商品，即可查看该商品的详情。

> **课外扩展：**
>
> 　　加入购物车时，可以给用户提供口味、数量等的选择，用户体验更好，也更加符合实际应用。

第 5 章
购物车界面

学习目标

通过购物车界面的实战学习,学生能够分析程序的基本功能,利用所学知识完成界面搭建,具体目标如下。

(1)掌握 ListView 组件、BaseAdapter 的使用方法,掌握 ListView 组件的优化方法。

(2)掌握 GSON 解析 JSON 数据的使用方法。

(3)掌握 Android 客户端与服务器端通信的方法。

5.1 任务描述

在商品详情界面可以单击【加入购物车】按钮,每次添加商品数量为1,添加成功(见图5-1)后,可以在购物车界面查看添加的商品(见图5-2)。

图 5-1 添加购物车成功效果图

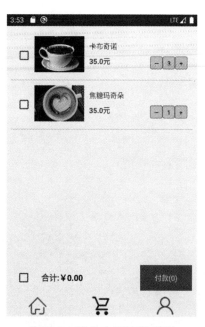

图 5-2 购物车界面效果图

5.2 相关知识

◆ 5.2.1 ListView 简介

在 Android 开发中,ListView 是一个比较常用的组件,它以列表的形式展示数据内容,并且能够根据屏幕的高度来显示滚动条自适应高度。

在为 ListView 组件添加数据时常会用到适配器。

◆ 5.2.2 优化 ListView 组件

在使用 ListView 展示数据时应该创建对应的 Item 来展示每条数据,如果展示的数据有成千上万条,那么就需要创建成千上万个 Item,这样会大大增加内存的消耗,甚至会由于内存溢出而导致程序崩溃。为了防止这种情况出现,需要对 ListView 进行两步优化。

1. 复用 covertView

在 ListView 第一次展示时,系统会根据屏幕的高度和 Item 的高度创建一定数量的 convertView。当滑动 ListView 时,顶部的 Item 会滑出屏幕并释放所有使用的 convertView,底部新的数据会进入屏幕进行展示,这时新的数据会使用顶部滑出 Item 的 convertView,从而在整个 ListView 展示数据的过程中使用固定数量的 convertView,避免了每次创建新的 Item 而消耗大量内存。

2. 使用 ViewHolder 类

复用 convertView 还不够，需要再使用 ViewHolder 类。

在加载 Item 布局时，会使用 findViewById（）方法找到 Item 布局中的各个控件，在每次加载新的 Item 数据时都会进行控件寻找，这样就会产生耗时操作，为了进一步优化 ListView 减少耗时操作，可以将要加载的子 View 放在 ViewHolder 类中，当第一次创建 convertView 时将这些控件找出，在第二次复用 convertView 时就可以直接通过 convertView 的 getTag（）方法获得这些控件。

使用 GSON 解析 JSON 数据。

GSON 由谷歌提供，使用 GSON 可以将 JSON 数据解析成实体类对象集合。使用 GSON 之前，需要创建 JSON 数据对应的实体类。需要注意的是，实体类中的成员名称要与 JSON 数据的 key 值一致。

5.3 具体步骤

◆ 5.3.1 使用 NavicatLite 新建购物车（shoppingcart）表

打开 NavicatLite，在 coffeedb 数据库下新建购物车（shoppingcart）表，表格字段设计如表 5-1 所示。

表 5-1 购物车（shoppingcart）表

字段名	字段类型	是否为主键	备注
cartId	int(11)	是（自增）	购物车编号
nums	int(11)		商品数量
checked	int(1)		商品是否选中，设置默认值为 0
mobile	varchare(255)	外键	手机号，不允许为空
goodsId	varchare(255)	外键	商品编号，不允许为空

◆ 5.3.2 处理添加购物车的 Servlet

1. 处理添加购物车的 Servlet 的作用

处理添加购物车的 Servlet 只是前端控制器，它的作用有以下三个。

（1）获取客户端发送的请求参数。

（2）处理用户请求。

（3）根据处理结果生成输出。

2. 添加购物车的接口

（1）请求地址：/SmartProductWeb/Android/AddToShoppingCartServlet。

（2）请求方法：post。

（3）请求头：charset=utf-8。

(4)返回头：text/html，charset=utf-8。

(5)请求实例：goodsId="1"&&nums="1"，携带登录状态。

(6)返回实例："success"字符串或"failed"字符串。

第 1 步：创建处理商品信息的 Servlet，选中 "servlets" 包，单击右键，选择【new】→【class】，创建 AddToShoppingCartServlet 类，如图 5-3 所示。

图 5-3　创建 AddToShoppingCartServlet 类

第 2 步：配置 AddToShoppingCartServlet 的访问路径。

将 AddToShoppingCartServlet 添加到 web.xml 中的 <web-app> 节点下，客户端可以通过 "IP 地址 :8080/SmartProductWeb/Android/AddToShoppingCartServlet" 的地址访问到该 Servlet，添加内容如下：

```
1   <servlet>
2       <servlet-name> AddToShoppingCartServlet </servlet-name>
3       <servlet-class>servlets. AddToShoppingCartServlet </servlet-class>
4   </servlet>
5   <servlet-mapping>
6       <servlet-name> AddToShoppingCartServlet</servlet-name>
7       <url-pattern>/Android/AddToShoppingCartServlet</url-pattern>
8   </servlet-mapping>
```

第 3 步：完成 AddToShoppingCartServlet 的功能。

获得要添加的商品 id，如果购物车表中已经存在该商品记录，则商品数量加 1，其他字段不变，不再增加新记录，如果购物车表中不存在该商品，则在购物车表中增加该条记录。

如果加入购物车成功，则返回给客户端字符串 "success"，否则返回给客户端字符串 "failed"。

/SmartProductWeb/src/servlets/ AddToShoppingCartServlet.java 代码如下：

```
1   package servlets;
2
```

```java
3   import java.io.IOException;
4   import java.io.PrintWriter;
5   import java.sql.ResultSet;
6   import java.sql.SQLException;
7
8   import javax.servlet.ServletException;
9   import javax.servlet.http.HttpServlet;
10  import javax.servlet.http.HttpServletRequest;
11  import javax.servlet.http.HttpServletResponse;
12
13  import db.MyDatabase;
14
15  public class AddToShoppingCartServlet extends HttpServlet {
16      public AddToShoppingCartServlet() {
17  }
18  @Override
19  protected void doPost(HttpServletRequest req,HttpServletResponse resp) throws ServletException,IOException {
20      req.setCharacterEncoding("utf-8");
21      resp.setCharacterEncoding("utf-8");
22      resp.setContentType("text/html,charset=utf-8");
23      // 获取需要存入数据库的信息
24      String values[]=new String[3];
25      values[0]=req.getParameter("nums");
26      values[1]=(String)req.getSession(false).getAttribute("mobile");
27      values[2]=req.getParameter("goodsId");
28      // 将该用户添加的商品信息插入购物车表中
29      MyDatabase mydatabase=new MyDatabase();
30      String query_sql="select * from shoppingcart where goodsId=?";
31      String update_sql="update shoppingcart set nums=?where goodsId=?";
32      String insert_sql="insert into shoppingcart(nums,mobile,goodsId) values(?,?,?)";
33
34      ResultSet rset=mydatabase.getSelectAll(query_sql,values[2]);
35      PrintWriter out=resp.getWriter();
36      int result=0;
37          try {
38              if(rset.next()) {
39                  int cnums=rset.getInt(2)+Integer.parseInt(values[0]);
```

```
40                          result=mydatabase.update(update_sql,cnums,
                            values[2]);
41                  }else {
42                          result=mydatabase.update(insert_sql,values);
43                  }
44          } catch (SQLException e) {
45              e.printStackTrace();
46          }
47          if(result==0) {
48              out.write("failed");
49          }else {
50              out.write("success");
51          }
52          mydatabase.closeDB();
53          out.flush();
54          out.close();
55      }
56  }
```

代码 20 ~ 22 行，调用 req.setCharacterEncoding（"utf-8"）方法设置请求内容编码格式为"utf-8"，调用 resp.setCharacterEncoding（"utf-8"）方法设置响应内容编码格式为"utf-8"，调用 resp.setContentType（"text/html,charset=utf-8"）方法设置响应内容为文本，编码格式为"utf-8"。

代码 25 行，调用 req.getParameter（"nums"）方法，从请求中获得 Android 客户端发来的需要添加的商品数量。

代码 26 行，调用 req.getSession（false）.getAttribute（"mobile"）方法，从 session 中取出该用户的手机号。

代码 27 行，调用 req.getParameter（"goodsId"）方法，从请求中获得 Android 客户端发来的需要添加的商品编号。

代码 38 ~ 43 行，根据 goodsId 在购物车表中的查询结果，如果有记录，说明该商品已经被添加了，则调用 mydatabase.update（update_sql, cnums, values[2]）方法修改该条记录中商品的购买数量；如果没有记录，则调用 mydatabase.update（insert_sql, values）方法，在购物车表中插入一条新的记录。

代码 47 ~ 51 行，如果插入失败，则调用 out.write（"failed"）方法，向 Android 客户端输出"failed"字符串；如果插入成功，则调用 out.write（"success"）方法，向 Android 客户端输出"success"字符串。

代码 52 行，调用自定义方法 mydatbase.closeDB（）来关闭数据库连接。

代码 53 ~ 54 行，将缓冲区数据输出，并关闭数据流。

5.3.3 完成商品界面的"加入购物车"功能

客户端通过AddToShoppingCartServlet将用户选购的咖啡添加到购物车表中,在客户端,需要提前获取到sessionid、添加商品的编号、购买数量。

因为使用session完成登录状态保持,用户的手机号存储在session中,所以Android客户端发送Servlet请求时,需要先将sessionid添加到请求头中,然后AddToShoppingCartServlet才能从session中取出该用户的手机号。

从共享参数文件login.xml中,取出用户登录时存储的sessionid值即可。

第1步:客户端新建"ShoppingCart"实体类。

在"entity"包下定义一个ShoppingCart类,如图5-4所示,在ShoppingCart类中设置购物车编号、商品购买数量、商品是否选中、商品信息、用户信息,且定义相应的方法。

图 5-4 添加 ShoppingCart 类

java\com\example\coffeedemo\entity\ShoppingCart.java 代码如下:

```
1   package com.example.coffeedemo.entity;
2
3   public class ShoppingCart {
4       private int cartId;
5       private int nums;
6       private int checked;// 商品是否被选中
7       private Goods goods;// 商品信息
8       private User user;// 用户信息
9
10      public int getCartId() {
11          return cartId;
12      }
13      public void setCartId(int cartId) {
14          this.cartId=cartId;
15      }
16      public int getNums() {
17          return nums;
18      }
```

```
19      public void setNums(int nums) {
20          this.nums=nums;
21      }
22      public int getChecked() {
23          return checked;
24      }
25      public void setChecked(int checked) {
26          this.checked=checked;
27      }
28      public Goods getGoods() {
29          return goods;
30      }
31      public void setGoods(Goods goods) {
32          this.goods=goods;
33      }
34      public User getUser() {
35          return user;
36      }
37      public void setUser(User user) {
38          this.user=user;
39      }
40  }
```

代码 4 ~ 8 行，是购物车实体类属性，读者可能会注意到此处属性与购物车表中的属性不一致。一般情况下，实体类属性设置与物理表字段设计是一致的，因为在购物车界面，除了需要商品编号外，还需要展示商品图片、商品名称等属性，为了显示数据方便，这里添加了商品对象作为属性，方便使用商品其他属性。

第 2 步：在商品详情界面 GoodsDetailActivity 类中，定义 addToCartByPost（String url, ShoppingCart sc）方法，连接 AddToShoppingCartServlet，将商品信息和用户信息添加到购物车表中。

如何知道是哪个用户添加了该商品呢？需要获得本地存储的 sessionid，将 sessionid 跟 post 请求一起发送到服务器，服务器从该 session 中取出用户手机号。在 GoodsDetailAcivity 类中定义 String getSessionid（）方法从 sharedPreferences 中取出 sessionid。代码如下：

```
1   public String getSessionid(){
2       SharedPreferences sp=getContext().getSharedPreferences("login",
        MODE_PRIVATE);
3       String msessionid=sp.getString("msessionid","");
4       return msessionid;
5   }
```

在 GoodsDetailAcivity 类中定义 addToCartByPost（String url, ShoppingCart sc, String msessionid）方法，代码如下：

```
1   public void addToCartByPost(String url,ShoppingCart sc,String
    msessionid){
2       RequestParams params=new RequestParams();
3       params.put("goodsId",sc.getGoods().getGoodsId());
4       params.put("nums",sc.getNums());
5       //实例化 AsyncHttpClient 对象
6       AsyncHttpClient client=new AsyncHttpClient();
7       client.addHeader("cookie","JSESSIONID="+msessionid);
8       //调用 post()方法连接网络
9       client.post(url,params,new TextHttpResponseHandler() {
10          @Override
11          public void onFailure(int i,Header[] headers,String s,
            Throwable throwable) {
12              Toast.makeText(GoodsDetailActivity.this," 连接网络失败 ",
                Toast.LENGTH_SHORT).show();
13          }
14          @Override
15          public void onSuccess(int i,Header[] headers,String s) {
16              if(s.equals("success")){
17                  Toast.makeText(GoodsDetailActivity.this," 添加购物车成
                    功 ",Toast.LENGTH_SHORT).show();
18              }else if(s.equals("failed")){
19                  Toast.makeText(GoodsDetailActivity.this," 添加购物车失
                    败 ",Toast.LENGTH_SHORT).show();
20              }
21          }
22      });
23  }
```

代码 3~4 行，此处使用购物车实体类设置需要存储的信息，调用已有的方法进行复制，不容易出错。

代码 7 行，调用 AsyncHttpClient 类的 client.addHeader（"cookie"，"JSESSIONID="+msessionid）方法，将 sessionid 一起发送给服务器端的 Servlet，保证 Android 客户端与服务器端的会话与登录时是同一个会话。

代码 16~20 行，连接网络成功后，判断 Servlet 返回的值，如果 Servlet 返回字符串"success"，则提示"添加购物车成功"，如果 Servlet 返回字符串"failed"，则提示"添

加购物车失败"。

第 3 步：添加 AddToShoppingCartServlet 的访问地址。

打开 res/strings.xml 文件，在 <resources> 节点下添加如下一行代码：

```
1  <string name="addtoshoppingcarturl">http://192.168.1.105:8080/
   SmartProductWeb/Android/AddToShoppingCartServlet</string>
```

第 4 步：在 4.3.12 节购物车详情界面 GoodsDetailActivity 类中添加代码，即在代码 34 行处添加如下代码：

```
1  ShoppingCart sc=new ShoppingCart();
2  sc.setNums(1);
3  sc.setGoods(goods);
4  addToCartByPost(getString(R.string.addtoshoppingcarturl),sc,getSessio
   nid());
```

代码 4 行，从 login.xml 文件中取出存储的 msessionid 值。

第 5 步：运行项目。

（1）运行服务器端"SmartProductWeb"，选中该项目，单击右键，选择【Run As】→【Run on sever】，出现如图 5-5 中内容即表示运行成功。

图 5-5　服务器启动成功效果图 1

（2）运行客户端 Android 应用程序，即可单击商品详情界面底部的【加入购物车】按钮，得到如图 5-1 所示效果图。

（3）"添加购物车成功"后，可以打开 NavicatLite 软件查看 shoppingcart 表，表中多了一条记录，如图 5-6 所示。

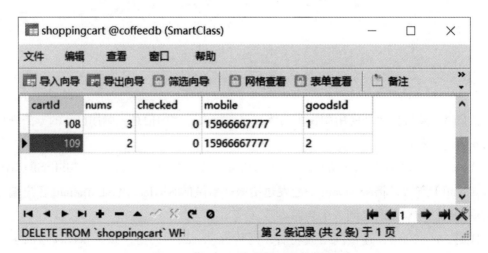

图 5-6　添加一种商品到购物车中

5.3.4 搭建购物车界面布局

购物车界面展示购物车信息，根据效果图（见图 5-2）完成界面布局，使用 <ListView> 展示购物车信息。

res\layout\fragment_cart.xml 代码如下：

```
1  <?xml version="1.0" encoding="utf-8"?>
2  <LinearLayout xmlns:android="http://schemas.android.com/apk/res/android"
3      android:layout_width="match_parent"
4      android:layout_height="match_parent"
5      android:orientation="vertical"
6      android:padding="10dp">
7      <ListView
8          android:id="@+id/lv"
9          android:layout_width="match_parent"
10         android:layout_height="wrap_content"
11         android:layout_weight="1"/>
12     <LinearLayout
13         android:layout_width="match_parent"
14         android:layout_height="50dp"
15         android:orientation="horizontal">
16         <LinearLayout
17             android:layout_width="0dp"
18             android:layout_height="match_parent"
19             android:layout_weight="2.5"
20             android:gravity="center_vertical"
21             android:orientation="horizontal">
22             <CheckBox
23                 android:id="@+id/all_chekbox"
24                 android:layout_width="wrap_content"
25                 android:layout_height="wrap_content"
26                 android:layout_gravity="center_vertical"
27                 android:layout_marginLeft="10dp"
28                 android:layout_marginRight="4dp"
29                 android:checkMark="?android:attr/listChoiceIndicatorMultiple"
30                 android:gravity="center"
31                 android:minHeight="64dp"
32                 android:paddingLeft="10dp"
33                 android:textAppearance="?android:attr/
```

```
                        textAppearanceLarge"
34                      android:visibility="visible"/>
35              <TextView
36                  android:layout_width="wrap_content"
37                  android:layout_height="wrap_content"
38                  android:layout_marginLeft="5dp"
39                  android:text=" 合计 :"
40                  android:textSize="16sp"
41                  android:textStyle="bold"/>
42              <TextView
43                  android:id="@+id/tv_total_price"
44                  android:layout_width="wrap_content"
45                  android:layout_height="wrap_content"
46                  android:text=" ￥0.00"
47                  android:textColor="#000"
48                  android:textSize="16sp"
49                  android:textStyle="bold"/>
50          </LinearLayout>
51          <TextView
52              android:id="@+id/tv_go_to_pay"
53              android:layout_width="0dp"
54              android:layout_height="match_parent"
55              android:layout_weight="1"
56              android:background="#E24146"
57              android:clickable="true"
58              android:gravity="center"
59              android:text=" 付款 (0)"
60              android:textColor="#FAFAFA"/>
61      </LinearLayout>
62  </LinearLayout>
```

◆ 5.3.5 处理购物车的 Servlet

1. 处理购物车的 Servlet 的作用

处理购物车的 Servlet 只是前端控制器，它的作用有以下三个。

（1）获取客户端发送的请求参数。

（2）处理用户请求。

（3）根据处理结果生成输出。

2. 购物车的接口

（1）请求地址：/SmartProductWeb/Android/ShoppingCartServlet。

（2）请求方法：post。

（3）请求头：charset=utf-8。

（4）返回头：text/html，charset=utf-8。

（5）请求实例：携带登录状态。

（6）返回实例：JSON 字符串。

代码如下：

```
1   [
2   {"nums":3,"checked":0,"goods":{"goodsImg":"http://192.168.1.105:8080/
    img/img_american.jpg","title":"卡布奇诺","price":35.0}},
3   {"nums":2,"checked":0,"goods":{"goodsImg":"http://192.168.1.105:8080/
    img/img_greentea.jpg","title":"焦糖玛奇朵","price":35.0}},
4   ]
```

第 1 步：创建处理商品信息的 Servlet，选中 "servlets" 包，单击右键，选择【new】→【class】，创建 ShoppingCartServlet 类，如图 5-7 所示。

图 5-7　创建 ShoppingCartServlet 类

第 2 步：配置 ShoppingCartServlet 的访问路径。

将 ShoppingCartServlet 添加到 web.xml 中的 <web-app> 节点下，客户端可以通过 "IP 地址:8080/SmartProductWeb/Android/ShoppingCartServlet" 的地址访问到该 Servlet，添加内容如下：

```
1   <servlet>
2       <servlet-name>ShoppingCartServlet</servlet-name>
3       <servlet-class>servlets.ShoppingCartServlet</servlet-class>
4   </servlet>
5   <servlet-mapping>
```

```
6        <servlet-name>ShoppingCartServlet</servlet-name>
7        <url-pattern>/Android/ShoppingCartServlet</url-pattern>
8    </servlet-mapping>
```

第 3 步：完成 ShoppingCartServlet 功能代码。

获取存储在 session 中的用户唯一标识—手机号，然后查询 shoppingcart 表，将该用户添加的所有记录以 JSON 格式返回给 Android 客户端。

/SmartProductWeb/src/servlets/ShoppingCartServlet.java 代码如下：

```
1   package servlets;
2   
3   import java.io.IOException;
4   import java.io.PrintWriter;
5   import java.sql.ResultSet;
6   import java.sql.SQLException;
7   
8   import javax.servlet.ServletException;
9   import javax.servlet.http.HttpServlet;
10  import javax.servlet.http.HttpServletRequest;
11  import javax.servlet.http.HttpServletResponse;
12  
13  import com.google.gson.JsonArray;
14  import com.google.gson.JsonObject;
15  
16  import db.MyDatabase;
17  
18  public class ShoppingCartServlet extends HttpServlet {
19      private static final long serialVersionUID=1L;
20      public ShoppingCartServlet() {
21      }
22  
23      @Override
24      protected void doPost(HttpServletRequest req,HttpServletResponse
            resp) throws ServletException,IOException {
25          req.setCharacterEncoding("utf-8");
26          resp.setCharacterEncoding("utf-8");
27          resp.setContentType("text/html,charset=utf-8");
28  
29          String mobile=(String)req.getSession(false).getAttribute
                ("mobile");
```

```
30        MyDatabase mydatabase=new MyDatabase();
31        String query_sql="select g.goodsImg,g.title,g.price,sc.nums,
              sc.checked from goods g,shoppingcart sc where
              g.goodsId=sc.goodsId and sc.mobile=?";
32        ResultSet rSet=mydatabase.getSelectAll(query_sql,mobile);
33        JsonArray mjsonarray=new JsonArray();
34        try {
35              while(rSet.next()) {
36                  JsonObject jsonObj=new JsonObject();
37                  jsonObj.addProperty("nums",rSet.getInt(4));
38                  jsonObj.addProperty("checked",rSet.getInt(5));
39                  JsonObject jsonObjGoods=new JsonObject();
40                  jsonObjGoods.addProperty("goodsImg",rSet.
                      getString(1));
41                  jsonObjGoods.addProperty("title",rSet.
                      getString(2));
42                  jsonObjGoods.addProperty("price",rSet.
                      getFloat(3));
43                  jsonObj.add("goods",jsonObjGoods);
44                  mjsonarray.add(jsonObj);
45
46              }
47        } catch (SQLException e) {
48              e.printStackTrace();
49        }
50        mydatabase.closeDB();
51        resp.getWriter().print(mjsonarray.toString());
52   }
53 }
```

代码 25～27 行，调用 req.setCharacterEncoding（"utf-8"）方法设置请求内容编码格式为"utf-8"，调用 resp.setCharacterEncoding（"utf-8"）方法设置响应内容编码格式为"utf-8"，调用 resp.setContentType（"text/html,charset=utf-8"）方法设置响应内容为文本，编码格式为"utf-8"。

代码 29 行，调用 req.getSession（false）.getAttribute（"mobile"）方法，从 session 中取出用户登录成功后添加的手机号。

代码 31、32 行，以该用户手机号作为查询条件，从购物车表中查询该用户添加的所有记录。

客户端向服务器端发送 /SmartProductWeb/Android/ShoppingCartServlet 请求，如果成功

将看到如图 5-8 所示的 JSON 响应数据。

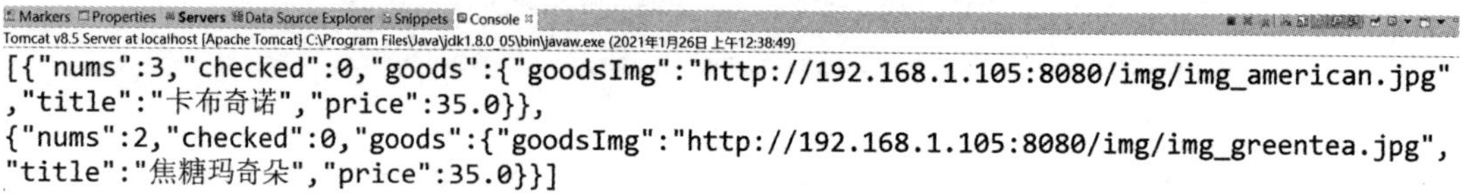

```
[{"nums":3,"checked":0,"goods":{"goodsImg":"http://192.168.1.105:8080/img/img_american.jpg"
,"title":"卡布奇诺","price":35.0}},
{"nums":2,"checked":0,"goods":{"goodsImg":"http://192.168.1.105:8080/img/img_greentea.jpg",
"title":"焦糖玛奇朵","price":35.0}}]
```

图 5-8　查询用户购物车信息的 JSON 响应

◆ **5.3.6　完成购物车界面功能逻辑实现**

购物车中的信息以列表形式展现，使用 <ListView> 组件完成，首先定义 <ListView> 每一条记录的布局，接着准备数据，创建适配器，最后绑定适配器。

第 1 步：定义 <ListView> 每一条记录的布局。

选中 "layout"，单击右键，选择【new】→【Layout resource file】，得到如图 5-9 所示界面，输入名称 "cart_lv_item"，然后单击【OK】按钮即可。

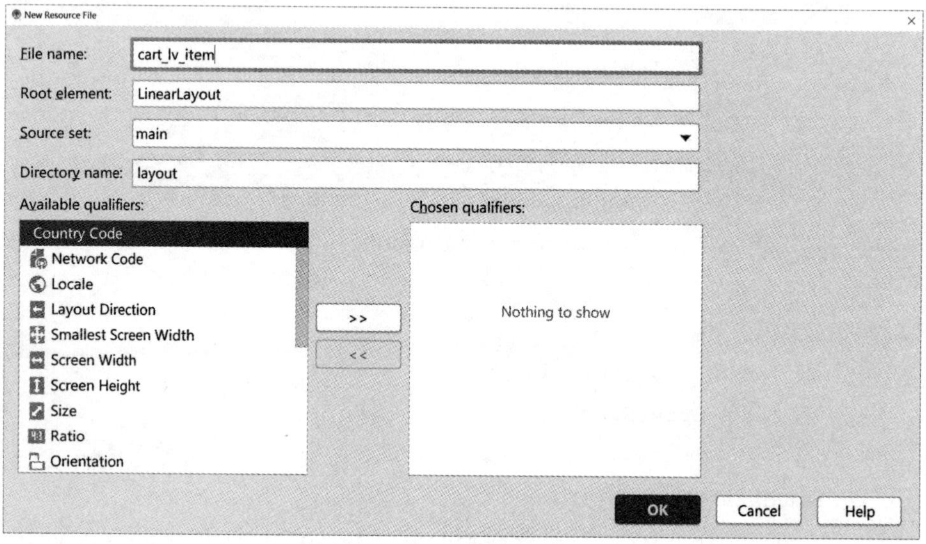

图 5-9　创建 <Listview> 的 Item 布局文件 cart_lv_item.xml

根据购物车界面效果图（见图 5-2）定义每一条 Item 的布局。

res\layout\cart_lv_item.xml 代码如下：

```xml
1  <?xml version="1.0" encoding="utf-8"?>
2  <LinearLayout xmlns:android="http://schemas.android.com/apk/res/
   android"
3      android:orientation="horizontal"
4      android:layout_width="match_parent"
5      android:layout_height="wrap_content"
6      android:padding="10dp">
7
8      <CheckBox
9          android:id="@+id/checkBox"
```

```xml
10          android:layout_width="wrap_content"
11          android:layout_height="wrap_content"
12          android:layout_gravity="center_vertical"
13          android:layout_marginRight="5dp"/>
14      <com.loopj.android.image.SmartImageView
15          android:id="@+id/sivIcon"
16          android:layout_width="100dp"
17          android:layout_height="70dp"
18          android:layout_marginBottom="5dp"
19          android:scaleType="centerCrop"
20          android:layout_marginRight="10dp"></com.loopj.android.image.SmartImageView>
21      <RelativeLayout
22          android:layout_width="match_parent"
23          android:layout_height="wrap_content">
24          <TextView
25              android:id="@+id/tv_goods_name"
26              android:layout_width="match_parent"
27              android:layout_height="wrap_content"
28              android:ellipsize="end"
29              android:maxLines="2"
30              android:textSize="14sp"
31              android:text="name"
32              android:layout_marginTop="10dp"/>
33          <TextView
34              android:id="@+id/tv_goods_price"
35              android:layout_width="wrap_content"
36              android:layout_height="wrap_content"
37              android:singleLine="true"
38              android:textSize="14sp"
39              android:textStyle="bold"
40              android:text="price"
41              android:layout_below="@id/tv_goods_name"
42              android:layout_marginTop="10dp"/>
43          <LinearLayout
44              android:layout_width="wrap_content"
45              android:layout_height="wrap_content"
46              android:orientation="horizontal"
47              android:layout_alignTop="@id/tv_goods_price"
48              android:layout_alignParentRight="true">
```

```
49              <TextView
50                  android:id="@+id/tv_reduce"
51                  android:layout_width="25dp"
52                  android:layout_height="25dp"
53                  android:background="@drawable/text_angle_border"
54                  android:gravity="center"
55                  android:text="-"
56                  android:textSize="12sp"
57                  android:textColor="#000"/>
58
59              <TextView
60                  android:id="@+id/tv_num"
61                  android:layout_width="25dp"
62                  android:layout_height="25dp"
63                  android:background="@drawable/text_angle_border"
64                  android:gravity="center"
65                  android:singleLine="true"
66                  android:text="1"
67                  android:textSize="12sp"
68                  android:textColor="#000"/>
69
70              <TextView
71                  android:id="@+id/tv_add"
72                  android:layout_width="25dp"
73                  android:layout_height="25dp"
74                  android:background="@drawable/text_angle_border"
75                  android:gravity="center"
76                  android:text="+"
77                  android:textSize="12sp"
78                  android:textColor="#000"/>
79          </LinearLayout>
80      </RelativeLayout>
81
82  </LinearLayout>
```

第 2 步：添加 ShoppingCartServlet 的访问地址。

打开 res/strings.xml 文件，在 <resources> 节点下添加如下一行代码：

```
1   <string name="shoppingcarturl">http://192.168.1.105:8080/
    SmartProductWeb/Android/ShoppingCartServlet</string>
```

第 3 步：创建 ShoppingCartAdapter 适配器。

购物车界面不仅仅是使用 <ListView> 完成购物车信息的展示,而且还需要给 Item 添加事件,如增加、减少商品数量,选中想要购买的商品等,下面就完成常见的购物车界面功能。

(1)创建适配器成员变量。

```
1   // 上下文
2   private Context myContext;
3   // 集合,存放购物车实体类数据
4   private List<ShoppingCart> mylist;
5   // 设置接口
6   private View.OnClickListener onAddNum;   // 增加商品数量接口
7   private View.OnClickListener onSubNum;   // 减少商品数量接口
8   private View.OnClickListener onCheck;    // 选中商品接口
9   private List<ShoppingCart> mylist;
```

(2)创建构造方法。

```
1   public ShoppingCartAdapter(Context context,List<ShoppingCart> shoppingCartList) {
2       myContext=context;
3       mylist=shoppingCartList;
4   }
```

(3)创建接口方法。

```
1   // 创建增加、减少商品的接口方法
2   public void setOnAddNum(View.OnClickListener onAddNum){
3       this.onAddNum=onAddNum;
4   }
5   public void setOnSubNum(View.OnClickListener onSubNum){
6       this.onSubNum=onSubNum;
7   }
8   // 选中商品的接口方法
9   public void setOnCheck(View.OnClickListener onCheck){
10      this.onCheck=onCheck;
11  }
```

(4)除了 getView()方法外,重写自定义适配器的其他三个方法的代码如下:

```
1   public int getCount() {
2       return mylist.size();
3   }
4   public Object getItem(int position) {
5       return position;
```

```
6     }
7     public long getItemId(int position) {
8         return position;
9     }
```

（5）定义 ViewHolder 内部类。

定义 ViewHolder 内部类，减少 Item 加载数据时控件的寻找，优化 ListView。

```
1   //定义内部类 ViewHolder
2   private static class ViewHolder{
3   //商品名称、数量、单价、图片
4   private TextView item_product_name;
5   private TextView item_product_num;
6   private TextView item_product_price;
7   private SmartImageView item_product_siv;
8
9   //增加、减少商品数量按钮
10  private TextView item_btn_add;
11  private TextView item_btn_sub;
12
13  //单选框
14  private CheckBox item_chekcbox;
15  }
```

（6）定义 getView（）方法。

```
1   public View getView(int position,View convertView,ViewGroup parent) {
2       ViewHolder holder;
3       /*view 重用 */
4       if(convertView==null){
5           convertView=
6           LayoutInflater.from(myContext).inflate(R.layout.cart_lv_item,
                parent,false);
7           holder=new ViewHolder();
8           holder.item_product_siv=(SmartImageView)convertView.findViewById
                (R.id.sivIcon);
9           holder.item_product_name=(TextView)convertView.findViewById(R.id.
                tv_goods_name);
10          holder.item_product_price=(TextView)convertView.findViewById(R.id.
                tv_goods_price);
11          holder.item_product_num=(TextView)convertView.findViewById(R.id.
                tv_num);
```

```
12      holder.item_chekcbox=(CheckBox)convertView.findViewById(R.id.
        checkBox);
13
14      // 设置增加、减少商品的接口回调
15      holder.item_btn_add=(TextView) convertView.findViewById(R.id.tv_add);
16      holder.item_btn_add.setOnClickListener(onAddNum);
17      holder.item_btn_sub=(TextView) convertView.findViewById(R.id.tv_
        reduce);
18      holder.item_btn_sub.setOnClickListener(onSubNum);
19
20      // 设置选中商品接口回调
21      holder.item_chekcbox.setOnClickListener(onCheck);
22      convertView.setTag(holder);
23      }else{
24      holder=(ViewHolder)convertView.getTag();
25      }
26      //SmartImageView 加载指定图片路径
27  holder.item_product_siv.setImageUrl(mylist.get(position).getGoods().
    getImg(),R.mipmap.ic_launcher,R.mipmap.ic_launcher);
28      holder.item_product_name.setText(mylist.get(position).getGoods().
        getTitle());
29      holder.item_product_price.setText(mylist.get(position).getGoods().
        getPrice()+" 元 ");
30      holder.item_product_num.setText(mylist.get(position).getNums()+"");
31      holder.item_chekcbox.setChecked(mylist.get(position).getChecked()
        ==0?false:true);
32      // 设置 Tag,用于判断用户当前单击的是哪一个列表项的按钮
33      holder.item_btn_add.setTag(position);
34      holder.item_btn_sub.setTag(position);
35      holder.item_chekcbox.setTag(position);
36
37      convertView.setTag(holder);
38
39      return convertView;
40  }
```

注意看加粗部分代码,一是实现了 convertView 的复用,二是使用 ViewHolder 类实现了 Item 数据加载时控件的复用。

代码 4～6 行,判断 convertView 是否为空,如果为空,则调用 LayoutInflater.from( ).infate( )方法引入自定的 Item 布局;如果不为空,则调用 convertView.getTag( )方法获得已

存在的控件（代码24行）。

java\com\example\coffeedemo\ui\ShoppingCartAdapter.java 代码如下：

```
1    package com.example.coffeedemo.ui;
2
3    import android.content.Context;
4    import android.view.LayoutInflater;
5    import android.view.View;
6    import android.view.ViewGroup;
7    import android.widget.BaseAdapter;
8    import android.widget.CheckBox;
9    import android.widget.TextView;
10
11   import com.example.coffeedemo.R;
12   import com.example.coffeedemo.entity.ShoppingCart;
13   import com.loopj.android.image.SmartImageView;
14
15   import org.w3c.dom.Text;
16
17   import java.util.List;
18
19   public class ShoppingCartAdapter extends BaseAdapter {
20       private Context myContext;
21       private List<ShoppingCart> mylist;
22       //设置接口
23       private View.OnClickListener onAddNum;   //增加商品数量接口
24       private View.OnClickListener onSubNum;   //减少商品数量接口
25       private View.OnClickListener onCheck;    //选中商品接口
26
27       public ShoppingCartAdapter(Context context,List<ShoppingCart> shoppingCartList) {
28           myContext=context;
29           mylist=shoppingCartList;
30       }
31       public int getCount() {
32           return mylist.size();
33       }
34       public Object getItem(int position) {
35           return position;
36       }
```

```java
37    public long getItemId(int position) {
38        return position;
39    }
40    public View getView(int position,View convertView,ViewGroup parent) {
41        ViewHolder holder;
42        /*View 重用 */
43        if(convertView==null){
44            convertView=LayoutInflater.from(myContext).inflate(R.layout.cart_lv_item,parent,false);
45            holder=new ViewHolder();
46            holder.item_product_siv=(SmartImageView)convertView.findViewById(R.id.sivIcon);
47            holder.item_product_name=(TextView)convertView.findViewById(R.id.tv_goods_name);
48            holder.item_product_price=(TextView)convertView.findViewById(R.id.tv_goods_price);
49            holder.item_product_num=(TextView)convertView.findViewById(R.id.tv_num);
50            holder.item_chekcbox=(CheckBox)convertView.findViewById(R.id.checkBox);
51
52            // 设置增加、减少商品的接口回调
53            holder.item_btn_add=(TextView) convertView.findViewById(R.id.tv_add);
54            holder.item_btn_add.setOnClickListener(onAddNum);
55            holder.item_btn_sub=(TextView) convertView.findViewById(R.id.tv_reduce);
56            holder.item_btn_sub.setOnClickListener(onSubNum);
57
58            // 设置选中商品接口回调
59            holder.item_chekcbox.setOnClickListener(onCheck);
60
61            convertView.setTag(holder);
62        }else{
63            holder=(ViewHolder)convertView.getTag();
64        }
65        //SmartImageView 加载指定图片路径
66        holder.item_product_siv.setImageUrl(mylist.get(position).getGoods().getImg(),R.mipmap.ic_launcher,R.mipmap.ic_
```

```
                launcher);
67              holder.item_product_name.setText(mylist.get(position).
                getGoods().getTitle());
68              holder.item_product_price.setText(mylist.get(position).
                getGoods().getPrice()+"元");
69              holder.item_product_num.setText(mylist.get(position).
                getNums()+"");
70              holder.item_chekcbox.setChecked(mylist.get(position).
                getChecked()==0?false:true);
71              // 设置Tag，用于判断用户当前单击的是哪一个列表项的按钮
72              holder.item_btn_add.setTag(position);
73              holder.item_btn_sub.setTag(position);
74              holder.item_chekcbox.setTag(position);
75
76              convertView.setTag(holder);
77
78              return convertView;
79          }
80      // 定义内部类 ViewHolder
81      private static class ViewHolder{
82          // 商品名称、数量、单价、图片
83          private TextView item_product_name;
84          private TextView item_product_num;
85          private TextView item_product_price;
86          private SmartImageView item_product_siv;
87
88          // 增加、减少商品数量按钮
89          private TextView item_btn_add;
90          private TextView item_btn_sub;
91          // 单选框
92          private CheckBox item_chekcbox;
93      }
94      // 创建增加、减少商品的接口方法
95      public void setOnAddNum(View.OnClickListener onAddNum){
96          this.onAddNum=onAddNum;
97      }
98      public void setOnSubNum(View.OnClickListener onSubNum){
99          this.onSubNum=onSubNum;
100     }
```

```
101        // 选中商品的接口方法
102        public void setOnCheck(View.OnClickListener onCheck){
103            this.onCheck=onCheck;
104        }
105    }
```

第 4 步：完成 Acivity 界面功能代码。

java\com\example\coffeedemo\fragment\CartFragment.java 代码如下：

```
1    package com.example.coffeedemo.fragment;
2
3    import android.content.Intent;
4    import android.content.SharedPreferences;
5    import android.os.Bundle;
6    import android.view.LayoutInflater;
7    import android.view.View;
8    import android.view.ViewGroup;
9    import android.widget.AdapterView;
10   import android.widget.CheckBox;
11   import android.widget.CompoundButton;
12   import android.widget.ListView;
13   import android.widget.RadioGroup;
14   import android.widget.TextView;
15   import android.widget.Toast;
16
17   import androidx.annotation.NonNull;
18   import androidx.annotation.Nullable;
19   import androidx.fragment.app.Fragment;
20
21   import com.example.coffeedemo.R;
22   import com.example.coffeedemo.entity.Goods;
23   import com.example.coffeedemo.entity.ShoppingCart;
24   import com.example.coffeedemo.ui.ShoppingCartAdapter;
25   import com.google.gson.Gson;
26   import com.google.gson.reflect.TypeToken;
27   import com.loopj.android.http.AsyncHttpClient;
28   import com.loopj.android.http.RequestParams;
29   import com.loopj.android.http.TextHttpResponseHandler;
30
31   import org.apache.http.Header;
32
33   import java.lang.reflect.Type;
```

```java
34  import java.util.ArrayList;
35  import java.util.List;
36
37  import static android.content.Context.MODE_PRIVATE;
38
39  public class CartFragment extends Fragment implements View.
    OnClickListener{
40      private ListView listView;
41      private ShoppingCart sc;
42      private ShoppingCartAdapter adapter;
43      private List<ShoppingCart> scList=new ArrayList<ShoppingCart>();
44      private double totalPrice=0.00;
45      private int totalCount=0;
46      private TextView mTvTotalPrice;
47      private TextView mTvGoToPay;
48      private CheckBox allCheckBox;
49
50      @Nullable
51      @Override
52      public View onCreateView(@NonNull LayoutInflater inflater,@
        Nullable ViewGroup container,@Nullable Bundle savedInstanceState)
        {
53          View view=inflater.inflate(R.layout.fragment_cart,container,
            false);
54          listView=(ListView) view.findViewById(R.id.lv);
55          mTvTotalPrice=(TextView)view.findViewById(R.id.tv_total_
            price);
56          mTvGoToPay=(TextView)view.findViewById(R.id.tv_go_to_pay);
57          allCheckBox=(CheckBox)view.findViewById(R.id.all_chekbox);
58          // 加载购物车内容
59          getShoppingCartFromServer(getString(R.string.shoppingcarturl),
            getSessionid());
60          // 商品全选、反选
61          allCheckBox.setOnCheckedChangeListener(new CompoundButton.
            OnCheckedChangeListener() {
62              @Override
63              public void onCheckedChanged(CompoundButton buttonView,
                boolean isChecked) {
64                  AllTheSelected(isChecked);
65              }
66          });
```

```java
67         return view;
68     }
69
70     public void getShoppingCartFromServer(String url,String msessionid) {
71         // 实例化 AsyncHttpClient 对象
72         AsyncHttpClient client=new AsyncHttpClient();
73         // post()请求时,携带登录状态
74         client.addHeader("cookie","JSESSIONID="+msessionid);
75         // 调用post()方法连接网络
76         client.post(url,new TextHttpResponseHandler() {
77             @Override
78             public void onFailure(int i,Header[] headers,String s,Throwable throwable) {
79                 Toast.makeText(getContext()," 连接网络失败 ",Toast.LENGTH_SHORT).show();
80             }
81
82             @Override
83             public void onSuccess(int i,Header[] headers,String s) {
84                 scList=parseJson(s);
85                 if(scList==null) {
86                     Toast.makeText(getContext()," 解析失败 ",Toast.LENGTH_SHORT).show();
87                 } else{
88                     // 创建适配器,加载 ListView 数据
89                     adapter=new ShoppingCartAdapter(getContext(),scList);
90                     listView.setAdapter(adapter);
91                     // 执行增加、减少商品数量的按钮单击事件接口回调
92                     adapter.setOnAddNum(CartFragment.this);
93                     adapter.setOnSubNum(CartFragment.this);
94                     adapter.setOnCheck(CartFragment.this);
95                 }
96             }
97         });
98     }
99
100    public List<ShoppingCart> parseJson(String json) {
101        // 使用 GSON 解析 JSON 数据
102        Gson gson=new Gson();
```

```java
103          // 创建一个 TypeToken 的匿名子类对象，并调用对象的 getType() 方法
104          Type listType=new TypeToken<List<ShoppingCart>>() {
105          }.getType();
106
107          List<ShoppingCart> scList=gson.fromJson(json,listType);
108          return scList;
109      }
110      public String getSessionid(){
111          SharedPreferences sp=getContext().
                 getSharedPreferences("login",MODE_PRIVATE);
112          String msessionid=sp.getString("msessionid","");
113          return msessionid;
114      }
115
116      //控制价格展示
117      private void priceContro() {
118          totalCount=0;
119          totalPrice=0.00;
120          for (int i=0; i<scList.size(); i++) {
121              if(scList.get(i).getChecked()==1) {
122                  totalCount=totalCount+scList.get(i).getNums();
123                  double goodsPrice=scList.get(i).getNums()*Double.
                         valueOf(scList.get(i).getGoods().getPrice());
124                  totalPrice=totalPrice+goodsPrice;
125              }
126          }
127          mTvTotalPrice.setText(" ￥ "+totalPrice);
128          mTvGoToPay.setText(" 付款 ("+totalCount+")");
129
130      }
131      /**
132       * 做全选和反选
133       */
134      private void AllTheSelected(Boolean aBoolean) {
135          if(aBoolean){
136              for (int i=0; i<scList.size(); i++) {
137                  scList.get(i).setChecked(1);
138              }
139          }else{
140              for (int i=0; i<scList.size(); i++) {
141                  scList.get(i).setChecked(0);
```

```
142                }
143            }
144        adapter.notifyDataSetChanged();
145        priceContro();
146    }
147
148    @Override
149    public void onClick(View v) {
150        Object tag=v.getTag();
151        switch (v.getId()){
152            case R.id.tv_add: // 单击添加数量按钮，执行相应的处理
153                // 获取 Adapter 中设置的 Tag
154                if (tag!=null&&tag instanceof Integer) {
                    // 解决问题：如何知道单击的按钮属于哪一个列表项中，通过 Tag
                    // 的 position
155                    int position=(Integer) tag;
156                    // 更改集合的数据
157                    int num=scList.get(position).getNums();
158                    num++;
159                    scList.get(position).setNums(num); // 修改集合中商品
                    // 数量
160                    // 解决问题：单击某个按钮时，如果列表项所需的数据改变了，
                    // 则应更新 UI
161                    adapter.notifyDataSetChanged();
162                }
163                break;
164            case R.id.tv_reduce: // 单击减少数量按钮，执行相应的处理
165                // 获取 Adapter 中设置的 Tag
166                if (tag!=null&&tag instanceof Integer) {
167                    int position=(Integer) tag;
168                    // 更改集合的数据
169                    int num=scList.get(position).getNums();
170                    if (num>0) {
171                        num--;
172                        scList.get(position).setNums(num);
                        // 修改集合中的商品数量
173                        adapter.notifyDataSetChanged();
174                    }
175                }
176                break;
177            case R.id.checkBox:// 单击选中商品按钮，执行相应的处理
```

```
178                    // 获取 Adapter 中设置的 Tag
179                    if (tag!=null&&tag instanceof Integer) {
180                        int position=(Integer) tag;
181                        // 更改集合的数据
182                        int checked=scList.get(position).getChecked();
183                        if (checked==0) {// 修改集合中商品选中状态
184                            scList.get(position).setChecked(1);
185                        }else{
186                            scList.get(position).setChecked(0);
187                        }
188                        adapter.notifyDataSetChanged();
189                        priceContro();
190                    }
191                    break;
192                }
193            }
194    }
```

第 5 步：运行项目。

（1）运行服务器端"SmartProductWeb"，选中该项目，单击右键，选择【Run As】→【Run on sever】，出现如图 5-10 所示内容即运行成功。

图 5-10　服务器启动成功效果图 2

（2）运行客户端 Android 应用程序，可以得到如图 5-11 的效果图。

图 5-11　购物车界面效果图

课外扩展：

（1）购物车界面的信息修改（如商品数量、商品是否选中等信息）后，没有更新到 shoppingcart 表中，读者可以自行实现。

（2）可以扩展实现付款功能、订单功能。

◆ 5.3.7 权限控制

前面介绍时已经看到，服务器端程序在处理用户登录时，如果用户登录成功，则系统会把用户手机号放入 HTTPSession 中，方便系统跟踪用户的登录状态。

对 Android 客户端程序来说，由于 Android 客户端采用了 AsyncHttpClient 来发送请求，获取响应，因此 HttpClient 会自动维护与服务器之间的登录状态，在有效时间内，服务器程序可以跟踪到 Android 客户端的登录状态。

本系统要求只有登录用户才能使用系统功能，因此程序需考虑在服务器端使用 Filter 进行控制，Filter 只要拦截匹配 /android/* 的 URL 即可。

/SmartProductWeb/src/servlets/Authority.java 代码如下：

```java
1   package servlets;
2
3   import java.io.IOException;
4
5   import javax.servlet.Filter;
6   import javax.servlet.FilterChain;
7   import javax.servlet.FilterConfig;
8   import javax.servlet.ServletException;
9   import javax.servlet.ServletRequest;
10  import javax.servlet.ServletResponse;
11  import javax.servlet.http.HttpServletRequest;
12  import javax.servlet.http.HttpSession;
13
14  public class Authority implements Filter{
15
16      public void init(FilterConfig arg0) throws ServletException {
17      }
18      public void doFilter(ServletRequest request,ServletResponse
        response,FilterChain chain)
19              throws IOException,ServletException {
20          HttpServletRequest hrequest=(HttpServletRequest) request;
```

```
21          // 获取 HttpSession 对象
22          HttpSession session=hrequest.getSession(true);
23          String mobile=(String)session.getAttribute("mobile");
24
25          // 如果用户已经登录或用户正在登录
26          if((mobile!=null&&mobile.length()>0)|| hrequest.
            getRequestURI().endsWith("/LoginServlet")
27
28          || hrequest.getRequestURI().endsWith("/RegistServlet")) {
29              //" 放行 " 请求
30              chain.doFilter(request,response);
31          }else {
32              response.setContentType("text/html;charset=utf-8");
33              // 生成错误提示
34              response.getWriter().println(" 您还没有登录系统，请先登录
                系统！ ");
35          }
36      }
37      public void destroy() {
38      }
39  }
```

在 web.xml 文件中添加 Filter 的配置信息，代码如下：

```
1  <filter>
2      <filter-name>Authority</filter-name>
3      <filter-class>servlets.Authority</filter-class>
4  </filter>
5  <filter-mapping>
6      <filter-name>Authority</filter-name>
7      <url-pattern>/android/*</url-pattern>
8  </filter-mapping>
```

代码 7 行，表示 Filter 将拦截的匹配 /android/* 的 URL 地址。

第6章 我的界面

学习目标

通过"我的界面"的实战学习,学生能够分析程序的基本功能,利用所学知识完成界面搭建,具体目标如下。

(1) 学会自定义下画线。

(2) 学会删除共享参数文件中内容。

(3) 学会关闭栈中所有 Activty。

6.1 任务描述

根据效果图搭建界面,在我的界面中只实现了"退出登录"功能,如图 6-1 所示。

图 6-1　我的界面效果图

6.2 相关知识

(1) Android 中如果要实现下画线,需要借助 shape 标签来完成。

(2) SharedPreferences.Editor 对象,调用 remove(String key)方法,可以删除一条记录。

(3) Intent 在实现界面跳转时,调用 Intent 对象的 setFlags(Intent.FLAG_ACTIVITY_CLEAR_TASK|Intent.FLAG_ACTIVITY_NEW_TASK)方法,可以关闭栈中所有的 Activity,将要启动的 Activity 置于栈顶。

6.3 具体步骤

◆ 6.3.1　搭建我的界面布局

根据效果图(见图 6-1)完成界面搭建。

第 1 步:创建下画线样式。

选中"drawable"文件夹,单击右键,选择【new】→【Drawable Resource File】,打开如图 6-2 所示窗口,输入 file name 为"shape_border_bottom",不带后缀名,单击【OK】按钮即可完成下画线的创建。

第 2 步:用 shape 实现下画线。

res\drawable\shape_border_bottom.xml 代码如下:

图 6-2　创建生成下画线样式的资源文件 shape_border_bottom.xml

```
1   <?xml version="1.0" encoding="utf-8"?>
2   <layer-list xmlns:android="http://schemas.android.com/apk/res/android">
3       <item
4           android:left="-2dp"
5           android:right="-2dp"
6           android:top="-2dp">
7           <shape>
8               <solid android:color="#00FFFFFF"/>
9               <stroke
10                  android:width="1px"
11                  android:color="#ccc"/>
12          </shape>
13      </item>
14  </layer-list>
```

第3步：定义"足迹""收藏夹""卡包"样式。

打开 res\values\styles.xml 文件，在 <resources> 节点下添加如下代码：

```
1   <style name="btnsInMine">
2       <item name="android:textSize">14dp</item>
3       <item name="android:gravity">center</item>
4       <item name="android:layout_weight">1</item>
5       <item name="android:padding">5dp</item>
6   </style>
```

第 4 步：修改 fragment_cart.xml 文件，完成我的界面布局。
res\layout\fragment_wode.xml 代码如下：

```
1   <?xml version="1.0" encoding="utf-8"?>
2   <LinearLayout xmlns:android="http://schemas.android.com/apk/res/android"
3       android:layout_width="match_parent"
4       android:layout_height="match_parent"
5       android:orientation="vertical">
6       <RelativeLayout
7           android:layout_width="match_parent"
8           android:layout_height="200dp"
9           android:background="#c4b7a4">
10          <LinearLayout
11              android:layout_width="match_parent"
12              android:layout_height="wrap_content"
13              android:orientation="horizontal"
14              android:layout_marginTop="40dp"
15              android:layout_marginLeft="20dp">
16              <ImageView
17                  android:layout_width="wrap_content"
18                  android:layout_height="wrap_content"
19                  android:src="@mipmap/ic_launcher"/>
20              <TextView
21                  android:layout_width="wrap_content"
22                  android:layout_height="wrap_content"
23                  android:text="Android学堂"
24                  android:layout_marginTop="20dp"
25                  android:layout_marginLeft="20dp"/>
26          </LinearLayout>
27          <LinearLayout
28              android:layout_width="match_parent"
29              android:layout_height="wrap_content"
30              android:orientation="horizontal"
31              android:layout_alignParentBottom="true">
32              <TextView
33                  android:layout_width="wrap_content"
34                  android:layout_height="wrap_content"
35                  android:text=" 足迹 "
36                  style="@style/btnsInMine"/>
```

```xml
37          <TextView
38              android:layout_width="wrap_content"
39              android:layout_height="wrap_content"
40              android:text=" 收藏夹 "
41              style="@style/btnsInMine"/>
42          <TextView
43              android:layout_width="wrap_content"
44              android:layout_height="wrap_content"
45              android:text=" 卡包 "
46              style="@style/btnsInMine"/>
47      </LinearLayout>
48  </RelativeLayout>
49  <LinearLayout
50      android:layout_width="match_parent"
51      android:layout_height="wrap_content"
52      android:orientation="horizontal"
53      android:padding="20dp"
54      android:background="@drawable/shape_border_bottom">
55      <TextView
56          android:layout_width="match_parent"
57          android:layout_height="wrap_content"
58          android:text=" 订单 "
59          android:layout_weight="1"
60          android:textSize="18dp"/>
61      <TextView
62          android:layout_width="wrap_content"
63          android:layout_height="wrap_content"
64          android:background="@drawable/ic_dot_right"/>
65  </LinearLayout>
66  <LinearLayout
67      android:id="@+id/btn_logout"
68      android:layout_width="match_parent"
69      android:layout_height="wrap_content"
70      android:orientation="horizontal"
71      android:padding="20dp"
72      android:background="@drawable/shape_border_bottom">
73      <TextView
74          android:layout_width="match_parent"
75          android:layout_height="wrap_content"
76          android:text=" 退出登录 "
```

```
77                    android:layout_weight="1"
78                    android:textSize="18dp"/>
79            <TextView
80                    android:layout_width="wrap_content"
81                    android:layout_height="wrap_content"
82                    android:background="@drawable/ic_dot_right"/>
83        </LinearLayout>
84  </LinearLayout>
```

代码 36、41、46 行，给组件添加 style= "@style/btnsInMine" 属性，加载定义好的样式。

代码 54、72 行，给组件添加 android:background= "@drawable/shape_border_bottom" 属性，加载下画线样式。

◆ 6.3.2　完成退出登录功能

实现退出登录分两步：首先删除用户登录成功时存储在本地的 sessionid，然后跳转到启动界面，关闭除了 MainActivity 之外的所有的 Activity。

java\com\example\coffeedemo\fragment\WodeFragment.java 代码如下：

```
1   package com.example.coffeedemo.fragment;
2
3   import android.content.Context;
4   import android.content.Intent;
5   import android.content.SharedPreferences;
6   import android.os.Bundle;
7   import android.view.LayoutInflater;
8   import android.view.View;
9   import android.view.ViewGroup;
10  import android.widget.LinearLayout;
11
12  import androidx.annotation.NonNull;
13  import androidx.annotation.Nullable;
14  import androidx.fragment.app.Fragment;
15
16  import com.example.coffeedemo.R;
17  import com.example.coffeedemo.activity.HomeActivity;
18  import com.example.coffeedemo.activity.MainActivity;
19
20  public class WodeFragment extends Fragment {
21      private LinearLayout btn_logout;
22
23      public View onCreateView(@NonNull LayoutInflater inflater,@
```

```
                Nullable ViewGroup container,@Nullable Bundle savedInstanceState){
24          View view=inflater.inflate(R.layout.fragment_wode,container,
            false);
25          btn_logout=(LinearLayout)view.findViewById(R.id.btn_logout);
26          btn_logout.setOnClickListener(new View.OnClickListener() {
27              public void onClick(View v) {
28                  // 清除用户登录时的 sessionid
29                  removeByKey("msessionid");
30                  // 跳转到 MainActivity,清除栈中其他 Activity
31                  navigateToWidthFlag(MainActivity.class,
32                  Intent.FLAG_ACTIVITY_CLEAR_TASK|Intent.FLAG_ACTIVITY_
                    NEW_TASK);
33              }
34          });
35          return view;
36      }
37      public void removeByKey(String key){
38          SharedPreferences sp=
39          getContext().getSharedPreferences("login",Context.MODE_
            PRIVATE);
40          SharedPreferences.Editor editor=sp.edit();
41          editor.remove(key);
42          editor.commit();
43      }
44      public void navigateToWidthFlag(Class cls,int flags){
45          Intent intent=new Intent(getActivity(),cls);
46          intent.setFlags(flags);
47          startActivity(intent);
48      }
49  }
```

代码 37 ~ 43 行，自定义方法 removeByKey（String key），删除用户登录成功时存储在共享参数文件中的 sessionid 值。

代码 44 ~ 48 行，自定义方法 navigateToWidthFlag（Class cls, int flags），实现 Activity 的跳转，并设置 flag。

代码 31 行，调用该方法 navigateToWidthFlag（MainActivity.class,Intent.FLAG_ACTIVITY_CLEAR_TASK|Intent.FLAG_ACTIVITY_NEW_TASK），实现了跳转到 MainActivity，并关闭其他 Activity。

参考文献

REFERENCES

[1] 黑马程序员. Android 移动应用基础教程（Android Studio）（第 2 版）[M]. 北京：中国铁道出版社，2019.

[2] 黑马程序员. Android 移动开发基础案例教程 [M]. 北京：人民邮电出版社，2017.

[3] 张荣. Android 开发与应用 [M]. 北京：人民邮电出版社，2014.

[4] 郭霖. 第一行代码 Android（第 2 版）[M]. 北京：人民邮电出版社，2016.